THE BIOLOGICAL IMPORTANCE OF BILE SALTS

NORTH-HOLLAND RESEARCH MONOGRAPHS

FRONTIERS OF BIOLOGY

VOLUME 47

Under the General Editorship of

A. NEUBERGER

London

and

E. L. TATUM†

New York

NORTH-HOLLAND PUBLISHING COMPANY
AMSTERDAM · NEW YORK · OXFORD

THE BIOLOGICAL IMPORTANCE OF BILE SALTS

G. A. D. HASLEWOOD

Emiritus Professor of Biochemistry in the University of London

1978

NORTH-HOLLAND PUBLISHING COMPANY
AMSTERDAM · NEW YORK · OXFORD

©Elsevier/North-Holland Biomedical Press, 1978

All rights reserved. No part of this publication may be reproduced, stored in a retrieval system, or transmitted, in any form or by any means, electronic, mechanical, photocopying, recording or otherwise, without the prior permission of the copyright owner.

Series ISBN: 0 7204 7100 1
Volume ISBN: 0 7204 0662 5

PUBLISHERS:
NORTH-HOLLAND PUBLISHING COMPANY — P.O. BOX 211, AMSTERDAM
(THE NETHERLANDS)

SOLE DISTRIBUTORS FOR THE U.S.A. AND CANADA:
ELSEVIER NORTH-HOLLAND INC.
52 VANDERBILT AVENUE, NEW YORK, NY 10017

Library of Congress Cataloging in Publication Data

Haslewood, G A D
   The biological importance of bile salts.

   (Frontiers of biology; v. 47) (North-Holland research monographs)
   Includes Bibliography and index.
     1. Bile salts. 2. Bile salt metabolism. I. Title.
II. Series: Frontiers of biology (Amsterdam); v. 47
QP752.B54H37   591.1'32   78-14494
ISBN 0-7204-0662-5 (v. 47)
ISBN 0-7204-7100-1 (Series)

PRINTED IN THE NETHERLANDS

To Beth in gratitude and love

Editors' preface

The aim of the publication of this series of monographs, known under the collective title of *"Frontiers of Biology"*, is to present coherent and up-to-date views of the fundamental concepts which dominate modern biology.

Biology in its widest sense has made very great advances during the past decade, and the rate of progress has been steadily accelerating. Undoubtedly important factors in this acceleration have been the effective use by biologists of new techniques, including electron microscopy, isotopic labels, and a great variety of physical and chemical techniques, especially those with varying degrees of automation. In addition, scientists with partly physical or chemical backgrounds have become interested in the great variety of problems presented by living organisms. Most significant, however, increasing interest in and understanding of the biology of the cell, especially in regard to the molecular events involved in genetic phenomena and in metabolism and its control, have led to the recognition of patterns common to all forms of life from bacteria to man. These factors and unifying concepts have led to a situation in which the sharp boundaries between the various classical biological disciplines are rapidly disappearing.

Thus, while scientists are becoming increasingly specialized in their techniques, to an increasing extent they need an intellectual and conceptual approach on a wide and non-specialized basis. It is with these considerations and needs in mind that this series of monographs, *"Frontiers of Biology"* has been conceived.

The advances in various areas of biology, including microbiology, bio-chemistry, genetics, cytology, and cell structure and function in general will be presented by authors who have themselves contributed significantly to these developments. They will have, in this series, the opportunity of bringing together, from diverse sources, theories and experimental data, and of integrating these into a more general conceptual framework. It is unavoidable, and probably even desirable, that the special bias of the individual authors will become evident in their contributions. Scope will also be given for presentation of new and challenging ideas and hypotheses for which complete evidence is at present lacking. However, the main emphasis will be on fairly complete and objective presentation of the more important and more rapidly advancing aspects of biology. The level will be advanced, directed primarily to the needs of the graduate student and research worker.

Most monographs in this series will be in the range of 200—300 pages, but on occasion a collective work of major importance may be included somewhat exceeding this figure. The intent of the publishers is to bring out these books promptly and in fairly quick succession.

It is on the basis of all these various considerations that we welcome the opportunity of supporting the publication of the series *"Frontiers of Biology"* by North-Holland Publishing Company.

E. L. TATUM†
A. NEUBERGER, *Editors*

Author's preface

There is nothing like a little practical application for stimulating research and the discovery that eating a common bile acid could, in suitable cases, lead to the disappearance of gallstones has had an explosive effect on the output of studies on bile salts. Even before this discovery, medical interest in these substances was increasing, as witnessed by the three volumes of *Bile Acids*, edited by P. P. Nair and D. Kritchevsky (1971, 1973, 1976) and K. W. Heaton's *Bile Salts in Health and Disease* (1972). I quote from these excellent works in the present text, which is an attempt to summarise a professional lifetime's thought and experience with these (to me) fascinating steroids. The present book is no sense a second edition of my *Bile Salts* (1967), being entirely re-written and with emphasis mainly on biological and medical rather than chemical matters: I have not dealt in any detail with the chemistry and have confined myself almost entirely to natural products. Much of the text is of course factual; some is idiosyncratic and speculative. The references, too, are a personal choice selected partly for their general interest and partly to assign credit in important original discoveries. In these times of computer storage and retrieval, completeness seems an unecessary aim and may make reading difficult and dull.

It is a pleasure to acknowledge the remarkable patronage to me and many other workers on liver problems of Dr. Herbert Falk, who has generously used his resources and those of his Companies in Freiburg, West Germany, and Basle, Switzerland, to encourage frequent international gatherings in these pleasant cities, at which much exchange of information and ideas, both publicly and privately, has taken place.

Papers read at Falk Congresses are published and are referred to in this book.

Many scientific and medical colleagues have helped me in writing this book, by discussion and by supplying publications; I express my gratitude. My practical contributions would have been impossible but for the generosity of those who have collected and supplied samples of bile. I thank the following for permission to refer to unpublished work: Dr. K. E. Banister, Dr. T. Briggs, Professor M. L. Johnson, Dr. R. J. Jørgensen, Dr. S. Ikawa, Dr. R. S. Oldham, Dr. A. R. Tammar and Dr. L. Tökés.

I thank Drs. Ehrlinger and K. R. Porter and Professor R. H. Dowling and their publishers for permission to reproduce Figures 1.2, 1.3 and 5.2 respectively from their works.

The author and publishers acknowledge with gratitude the skilful assistance of the Medical Illustration and Photography Departments of Guy's Hospital and especially Miss S. Fisk for the formulae and Schemes and Mr. P. Elliott for Figure 1.

I especially thank Mrs. Lynn C. Lambden for her cheerful patience and skill in preparing the manuscript.

Contents

Editors' Preface	vii
Author's Preface	xi
Chapter 1. Nature and function of bile	1
1.1. General anatomy and physiology	3
1.2. General function	6
1.3. Nature of bile: general	6
1.4. Nature of bile: mucin and protein	7
1.5. Nature of bile: bile pigments	7
1.6. Nature of bile: phospholipids	8
1.7. Nature of bile: fats other than phospholipid	9
1.8. Function of bile salts: fat digestion and absorption	11
1.9. Functions of bile salts other than in fat absorption	13
1.10. Effect of bile salts on parasites	13
1.11. Bile salts in tissues other than liver	15
Chapter 2. Chemistry and methods of separation and analysis	19
2.1. Definitions	21
2.2. General nature of bile salts	21
2.3. Bile alcohol sulphates	22
2.4. Taurine conjugates of C_{27} acids	26
2.5. Taurine and glycine conjugates of C_{24} bile acids	28
2.6. Unconjugated bile acids acting as bile salts	32
2.7. Other conjugates	34
2.8. Methods of separation and analysis	35
2.9. Physical chemistry	39

Chapter 3. Biosynthesis and artifacts of the enterohepatic circulation   47
3.1. Biosynthesis: (a) C_{24} 5β bile acids   49
3.2. Biosynthesis: (b) C_{24} 5α bile acids   59
3.3. Biosynthesis: (c) C_{27} and C_{28} acids   61
3.4. Biosynthesis: (d) bile alcohols   62
3.5. Conjugation   66
3.6. Control of bile salt biosynthesis   67
3.7. Artifacts of the enterohepatic circulation   69

Chapter 4. Distribution of bile salts in the animal kingdom   77
4.1. Invertebrates   79
4.2. Vertebrates   80
    4.2.1. Cyclostomes   80
    4.2.2. Condrichthyans   82
    4.2.3. Osteichthyans   83
    4.2.4. Amphibians   93
    4.2.5. Reptiles   96
    4.2.6. Birds   102
    4.2.7. Mammals   103

Chapter 5. The importance to medicine of bile salts   119
5.1. Composition of human bile   121
5.2. Human bile salts (a) in bile   123
5.3. Human bile salts (b) in normal blood   125
5.4. Human bile salts (c) in normal urine   127
5.5. Human bile salts (d) in faeces   127
5.6. The solubility of cholesterol in bile   128
5.7. The gallstone problem   130
5.8. The bile salt "pool"   135
5.9. Bile salts in liver disease   136
5.10. Bile salts in biliary atresia   139
5.11. Other diseases involving bile salts   139
5.12. The breath test   141
5.13. Effect of diet on cholesterol balance   142

Chapter 6. General conclusions and speculations   151

Appendix. Bile salts in different animal forms   161
Addendum   183
Index   195

CHAPTER 1

Nature and function of bile

1.1. General anatomy and physiology

Bile is produced by the liver: all vertebrates so far investigated have both a liver and bile. At the lowest level of organisation amongst true vertebrates, the hagfishes (Myxinoidea) have a large gall-bladder and ample bile. In vertebrate animals below birds and mammals, I have never encountered a species lacking a gall-bladder but this organ has been lost in a number of birds and mammals, some or all of which have a sphincter at the junction of the bile duct with the small intestine which enables the flow of bile to be controlled to some extent. Fig. 1.1 shows diagrammatically the anatomical arrangement in man.

Fig. 1.1 shows that the bile is stored in the gall-bladder (from which it is released by contraction under stimulus from the peptide hormone cholecystokinin) and reaches the duodenum. In man, the sphincter of Oddi is present at the entrance of the bile duct into the duodenum, an entrance usually shared with that of the pancreatic duct. In other vertebrates, for example snakes, the gall-bladder and liver may be quite widely separated. Anatomical relationships different from those shown in Fig. 1.1, between liver, gall-bladder and bile ducts exist in various animal forms (e.g. Sobotka, 1937). The presence of an enterohepatic circulation means that the bile may normally contain substances modified by microorganisms living in the intestine. Dowling (1972), in reviewing this situation, pointed out that bile salts are probably the only compounds undergoing "bulk" enterohepatic circulation. An active transport mechanism for absorbing bile salts exists, mainly in the terminal ileum, but some passive non-ionic diffusion (especially of bile acids made by intestinal bacteria from

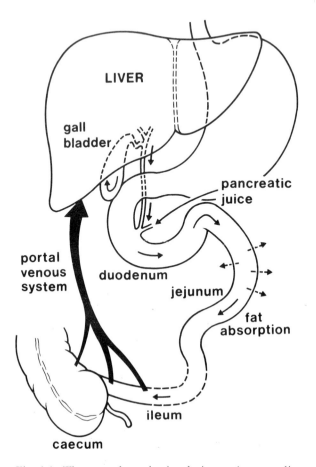

Fig. 1.1. The enterohepatic circulation as in man, diagrammatically sketched.

their conjugates) may occur throughout the small intestine and also from the caecum and large bowel. Dietschy (1968) has reviewed this aspect of bile physiology. Few have studied the enterohepatic circulation in animals other than birds and mammals.

The gall-bladder concentrates some constituents, for example bile pigments and bile salts, Na^+ and K^+, whilst causing the loss of others, e.g. Cl^- and HCO_3^-. The pH of its contents is usually less than that of hepatic bile. The properties of gall-bladder bile (isotonic to blood plasma, high concentrations of Na^+ and K^+ and low Cl^-) are much the same in all classes of vertebrates (Diamond, 1968). According to Rose et al. (1973): "The presence of a significant transmural potential difference in human gall-bladder suggests but does not prove that an electrogenic Na⁺ transport mechanism is operative".

Much of the bile is formed, in the species examined, by hepatocytes and excreted by these cells into bile canaliculi. In man, these are channels of diameter about 1 μm, closed at one end from the intercellular space by tight junctions and connected at the open ends with one another to form a continuous network. At their open ends, bile canaliculi drain into bile ductules which join to form bile ducts. The way in which a bile canaliculus is formed between adjacent hepatocytes, in man, is shown in Fig. 1.2, which is taken from the

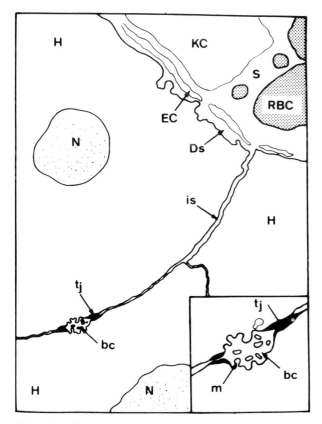

Fig. 1.2. Bile canaliculus in relation to hepatocytes and Kupfer cells in human liver (× 14,000). H, hepatocyte; N, nucleus; KC, Kupfer cell; RBC, red blood cell; EC, endothelial cell; S, liver sinusoid; is, intercellular space; Ds, space of Disse (perisinusoidal space); bc, bile calaniculus; tj, tight junction; m, microvilli. (Taken, with permission, from Ehrlinger and Dhumeaux, 1974).

review of Ehrlinger and Dhumeaux (1974) and is a drawing from an electron micrograph. The bile canaliculus contains microvilli (m, Fig. 1.2) and these perhaps play an important part in bile production The ductules and ducts are lined with an epithelium which could alter the composition of bile, for example by adding mucin, adding or subtracting water or ions such as bicarbonate. An excellent short description of the microscopic structures concerned with canalicular bile secretion is given by Schaffner (1975) and Rappaport (1975) has a clear picture of the circulatory system involved. Membrane receptors on the sinusoidal surfaces may control canalicular bile secretion "where one cell whispers to another something about bile formation" (Desmet, 1976). The canalicular membrane excretes bile salts by a carrier mechanism (Simon et al., 1977) and a similar system may be involved in uptake by liver cells from the portal circulation (Reichen et al., 1977).

It is generally believed that there is canalicular bile and also bile produced in the ducts themselves: man may normally produce about 450 ml of canalicular and about 150 ml of ductular bile every 24 h. Both canalicular and ductular bile secretion are divisible into a "bile-salt dependent" and a "bile-salt independent" phase. The latter type of secretion, quantitatively less important than the former, may be governed by the Na^+-K^+ ATP-ase transport system within the canalicular membranes which is much affected by the thyroid hormone (Ehrlinger and Dhumeaux, 1974).

1.2. General function

The bile can be regarded both as a digestive secretion, particularly concerned with the digestion and absorption from the intestine of fats, and also as a channel for the excretion of substances for one reason or another not suitable for elimination via the kidneys. Smith (1973) has reviewed this latter aspect of physiology as it concerns not only substances normally excreted but also drugs and other compounds foreign to the body.

1.3. Nature of bile: general

Bile itself contains water, mucin, protein, bile pigments, bile salts, phospholipids such as lecithin, neutral fats (e.g. cholesterol, glycerides), urea, Na^+, K^+, Ca^{2+}, Fe^{2+}, Cl^-, HCO_3^- and other inorganic ions and, in some mammals at least, the enzyme alkaline phosphatase. For the composition of human bile, see Chapter 5.

1.4. Nature of bile: mucin and protein

The mixture of mucopolysaccharide "molecules" constituting this substance is always present and the concentration seems also always to be greater in gall-bladder than in hepatic bile. Schrager (1970) gives a general review of the properties of gastrointestinal mucus; however, little is known about the particular nature of this material in bile. The matter ought to be more carefully studied, especially since some of the bile pigment is closely associated with the mucin fraction and not separable by ethanol precipitation or gel filtration (Bouchier and Cooperband, 1967; Gent et al., 1971).

Clausen et al. (1965) made a careful examination of the proteins of chicken bile and that from human patients with and without gallstones. Normal human bile contained about 20—50 μg/ml of both albumin and γ-G-globulin. Albumin was also found in half the human gallstones examined and was generally at a higher concentration in bile from patients with gallstones than in "normal" bile. The proteins found appeared in general to be the same as those present in blood serum. Later work is quoted by Espinosa (1976), who identified antigens in human gall-bladder bile.

1.5. Nature of bile: bile pigments

These compounds are generally held to be the excreted result of endogenous haemoglobin degradation in the reticuloendothelial system, especially liver and spleen. I have long believed that a substantial exogenous component derived from dietary blood pigments is present also and discuss this view in Chapter 5.

Chemically, the chief bile pigments are supposed to be bilirubin and biliverdin; there is good evidence that biliverdin from haem degradation is reduced in vivo to bilirubin, which the liver conjugates covalently chiefly with carbohydrate and to a small extent with sulphate before excretion into the bile. Although many biles are green in colour, little is known about the excretion of biliverdin itself, although it has been identified in fowl and amphibian bile (see Fevery et al., 1977).

There has been, and still is, controversy about the chemical nature of biliary bilirubin. Gordon et al. (1976) give a brief summary of the history of this argument and interpret their own work as showing that the principal pigment in human and dog bile is bilirubin diglucuronide. Fevery et al. (1977) found both glucuronide and glucoside conjugates in the bile and corresponding conjugating enzymes in the liver and kidney tissue of 11 animal species, including man. Mono- and diglucuronides were chief pigments in 10 mammals; in the domestic fowl "glucosides and glucuronide conjugates were of equal importance", but "probably represent only a small fraction of the total bile pigment". Renal conjugation seemed unimportant for most species, including man. Gordon et al. (1976) do not mention the mucin fraction and it is not possible to gather from their paper how they disposed of it. As mentioned above, other authors have not been able to separate some of the pigment from a macromolecular complex and the intriguing possibility remains that some bilirubin might be covalently linked to polysaccharide. Etter-Kjesaas and Kuenzle (1975) have isolated a polypeptide-bilirubin conjugate of molecular weight about 7,000 from human bile: the bilirubin is covalently linked to the polypeptide.

Kuenzle (1970) demonstrated that in human cholestatic bile, conjugates with numerous carbohydrates other than glucuronate are present; these are apparently pathological and occur in minor amounts in bile from healthy people. In spite of the work of Gordon et al. and others (e.g. Jansen and Billing, 1971; Heirwegh et al., 1975; Blumenthal et al., 1977), our knowledge of biliary bilirubin is very incomplete: it is agreed however, that in health no more than very small amounts of unconjugated bile pigment can be detected. Kuenzle (1970) suggested that some of his carbohydrate-bilirubin conjugates might have amphiphilic (detergent) properties; they are not thought, however, to be important in fat digestion or in solubilising lipid in the bile. So we can say with fair conviction that the bile is a channel for the excretion of bile pigments, which have no physiological function.

1.6. Nature of bile: phospholipids

The phospholipids identified in bile so far are chiefly lecithins. An example is the palmityl-oleyl-phosphatidyl choline (Formula 1.1) of human bile: bile salts probably regulate hepatic biosynthesis of such compounds (Marks et al., 1976).

$$\text{CH}_3\cdot(\text{CH}_2)_7\cdot\text{CH}=\text{CH}\cdot(\text{CH}_2)_7\cdot\text{CO}\cdot\text{O}\cdot\overset{2}{\text{C}}\text{H} \quad \begin{array}{l} {}^1\text{CH}_2\cdot\text{O}\cdot\text{CO}\cdot(\text{CH}_2)_{14}\cdot\text{CH}_3 \\ | \\ {}^3\text{CH}_2\cdot\text{O}\cdot\overset{\overset{\text{O}}{\|}}{\underset{\underset{\text{O}^-}{|}}{\text{P}}}\cdot\text{O}\cdot\text{CH}_2\cdot\text{CH}_2\cdot\overset{+}{\text{N}}(\text{CH}_3)_3 \end{array}$$

1.1 L-1-palmityl-2-oleyl-phosphatidyl choline a biliary lecithin.

Such zwitterionic substances have been described as swelling amphiphiles to indicate that, as well as having hydrophilic and lipophilic parts in one molecule, they are essentially insoluble in water but readily take it up and so swell. These properties no doubt account for the role of phospholipids in solubilising cholesterol and other fats in the bile, a subject discussed in Chapter 5. It seems probable, therefore, that biliary phospholipids are physiologically functional. In the small intestine, lecithins can be hydrolysed by pancreatic phospholipase A, for example, which would remove the 2-oleoyl group from 1.1, forming a lysolecithin with amphiphilic detergent properties. In some animal species, for example sheep, lecithins and lysolecithins are thought to be most important in fatty acid absorption (Harrison and Leat, 1975; Leat and Harrison, 1977; Lough and Smith, 1976).

1.7. Nature of bile: fats other than phospholipid

At one time it was supposed that bile contains a considerable amount of fatty acids, an idea perpetuated in Sobotka's (1937) classical book. More modern methods suggest that in human bile at least the concentration of these substances is small and is exceeded by that of monoglycerides (Table 5.1, p. 122). Nakayama and van der Linden (1970) found small amounts of diglyceride but no triglyceride in human gall-bladder bile.

It is probably true to say that the glycerides and fatty acids are excreted substances only and this may also be the case with cholesterol. This compound is of great medical interest and its solubility in bile is discussed in detail in Chapter 5. As far as is known at present, all biliary cholesterol is present as such; for this reason its concentration could easily be determined by precipitation as the very insoluble digitonide. Digitonin is unlikely to precipitate any other substance, except perhaps cholestanol (5α-cholestan-3β-ol) from bile. Few workers have used this method, however, and some major mistakes in estimating biliary cholesterol have been made because of a failure to recognise that the bile salts, usually present in much greater concentration than cholesterol, also respond in most of the colour reactions used for its estimation. Indeed, these colour reactions (the Liebermann-Burchard reaction and the like) are not particularly specific, being given by many terpenoid compounds and even (in my experience) by the purest samples of the saturated 5β-cholestan-3β-ol (coprosterol, coprostanol). Cholesterol can be concentrated by partition of diluted bile between light petroleum and aqueous ethanol or methanol: lipids including cholesterol will be almost entirely in the petroleum layer and bile salts in the aqueous.

Animal species differ greatly in the concentration of their biliary cholesterol, man normally having much greater amounts than most, if not all, other animals. How much biliary cholesterol is exogenous and how much derived from biosynthesis is an important question not yet satisfactorily settled.

Cholesterol is of course the biogenetic source of the bile salts (Chapter 3) and there is some evidence that the newly-biosynthesised fraction is used for this purpose. Indeed, some authors go further and say that newly-synthesized hepatic cholesterol is the precursor for both bile salts and the biliary cholesterol itself (e.g. Normann and Norum, 1976). Percy-Robb (1975), however, believes that "from a dynamic point of view, liver and plasma cholesterol can be considered as a single pool". Normann and Norum (1976) agree with an earlier view in saying "It is therefore probable that the liver production of bile constituents is so organised that secreted products derive from an anabolic pool and not from a catabolic pool". If by this slightly cryptic comment the authors are referring to all substances cleared from the blood by the liver into the bile, their opinion is clearly untrue, for example, of conjugated bile pigments, which can be removed by the liver from the general circulation on recovery from obstructive jaundice. On the other hand, if Normann and Norum are using "secreted" to mean physiologically functional, then I should not put biliary cholesterol into that category; on the contrary, it is often deleterious to man (Chapter 5). Furthermore feeding cholesterol, to many primates at least, results in a considerably increased output of this substance in the bile (e.g. Tanaka et al., 1976); it is hard to believe that this is the result of a stimulation of hepatic biosynthesis.

1.8. Function of bile salts: fat digestion and absorption

The bile is also to be regarded as an important digestive fluid and in this respect the bile salts are its principal physiologically active constituent (see, however, section 1.6). Present day views on non-ruminant mammalian digestion and absorption of long-chain triglyceride fat (the chief component of dietary "fat" in most species) are neatly summarised by Porter (1969) in Fig. 1.3.

Long chain triglycerides, emulsified by the detergent action of bile salts, are digested by pancreatic lipase mainly to 2-monoglycerides and fatty acids (Scheme 1.1).

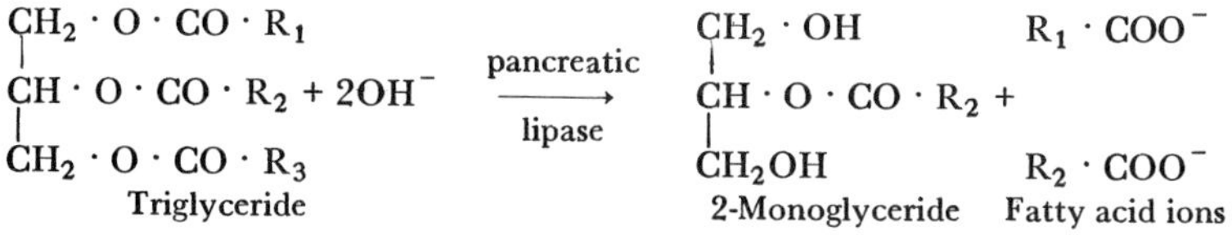

$$\begin{array}{c} CH_2 \cdot O \cdot CO \cdot R_1 \\ | \\ CH \cdot O \cdot CO \cdot R_2 + 2OH^- \\ | \\ CH_2 \cdot O \cdot CO \cdot R_3 \\ \text{Triglyceride} \end{array} \xrightarrow{\text{pancreatic lipase}} \begin{array}{c} CH_2 \cdot OH \\ | \\ CH \cdot O \cdot CO \cdot R_2 \\ | \\ CH_2 OH \\ \text{2-Monoglyceride} \end{array} + \begin{array}{c} R_1 \cdot COO^- \\ \\ \\ R_2 \cdot COO^- \\ \text{Fatty acid ions} \end{array}$$

Scheme 1.1. The chief digestive process for triglyceride fat

As might be expected, bile salts stimulate the action of intestinal lipases especially in the presence of their co-factors, co-lipases (e.g. Borgström, 1977). They also activate and protect cholesterol esterase derived from the pancreas (see Nair, 1976). As digestion proceeds, sub-division of particles increases until micelles incorporating fatty acid ions and monoglyceride and stabilised by bile salts are formed. It is not clear how these micelles now behave; the opinion expressed by Fig. 1.3 is that they shed their fatty acid ions and monoglycerides into the microvilli of intestinal cells of the duodenum and proximal jejunum, releasing the bile salts to continue to aid digestion of further fat. The micelles discussed here are negatively charged, as is the original bile salt-fat emulsion because, as shown in Chapter 2, the polar OH, COO$^-$ and SO$_3^-$ groups will lie in the aqueous medium whilst the molecules comprising the micelle adhere by "dissolving" in one another at their hydrophobic fatty parts, which likewise also dissolve in the fat during digestion. Cholesterol and other fats, including phospholipids, may be incorporated in the interior of such particles, finally forming "mixed micelles". This micellar theory of fat absorption is not at present contradicted; it is thought to express events with fatty acids having straight chains of 14 or more carbon atoms. As the chains are shortened more and more below C_{14}, fatty

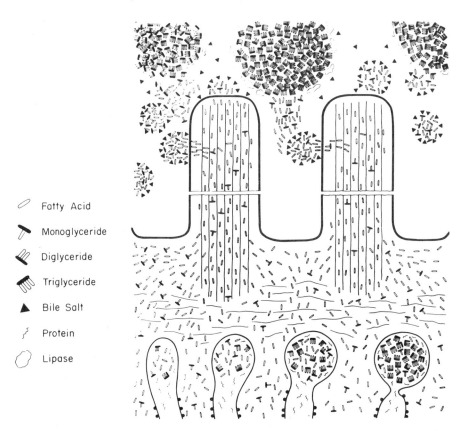

Fig. 1.3. The intraluminal digestion of triglyceride and the selective absorbtion of 2-monoglycerides and long-chain fatty acids from mixed micelles into the microvilli of a single epithelial intestinal cell, followed by re-synthesis of triglyceride within the smooth endoplasmic reticulum of the cell. (Taken, with permission, from Porter (1969)).

acids are increasingly absorbed directly into the portal circulation, passing without re-synthesis of triglycerides to the liver. Senior (1964) provides a summary of the evidence for these ideas and a discussion of the subsequent fate of re-synthesized fat.

1.9. Functions of bile salts other than in fat absorption

In addition to their function in the digestion and adequate absorption of triglyceride fat, cholesterol, fat-soluble vitamins and other lipids, it has been suggested that bile salts may also be important for the intestinal transport and intracellular metabolism of water-soluble nutrients such as glucose and amino acids. After investigating these suggestions, Pope et al. (1966) concluded that purified glycine or taurine conjugates of cholic acid did not affect a number of transport activities across isolated perfused rat jejunum. However deoxycholate (a common contaminent of commercial bile salt preparations) did act as an inhibitor of jejunal transport. It is unlikely that unconjugated deoxycholate would reach the bile, except perhaps in disease, so that the work of Pope et al. does not support the notion of a physiological role for bile salts in the metabolism of water-soluble constituents of the diet.

Nair (1976) reviews the action of bile salts on enzymes generally and attributes some effects to bile salt-protein binding, which may be covalent.

Javitt et al. (1973) reported that two normal intermediates in bile salt biosynthesis, namely 5β-cholestane-3α,7α-diol and 5β-cholestane-3α,7α,12α-triol (Scheme 3.1), enhanced the rate of porphyrin synthesis by cultured chick liver cells. The 5β-cholestane alcohols apparently did this by inducing δ-aminolaevulinate synthetase, the rate-limiting enzyme for haem biosynthesis. This finding, if confirmed, shows a link between two important liver functions and may help to explain over-production of porphyrins in liver disease, for example cirrhosis.

1.10. Effect of bile salts on parasites

Echinococcus granulosus, the cestode parasite responsible for hydatid disease in man, has a definitive host in which the organism can come to maturity and produce eggs which are then excreted and can undergo partial development as far as the protoscolex stage in an intermediate host, forming hydatid cysts often in the liver. The commonest definitive host is a canid such as a dog or fox and the intermediate host a herbivore or omnivore, for example sheep, deer, pig or man. Smyth and Haslewood (1963) found that dihydroxy types of bile salts in the intermediate host (Chapter 4) could prevent development of the parasite by lysis of its protoscolex stage. Thus, the parasites' complete development is possible only in animals such as carnivores with little dihydroxy bile salt. Perhaps, also, the taurocholate that is the chief bile salt in dogs has a positive effect on development of *E. granulosus*. In an analogous way, other intestinal parasites may be inhibited or stimulated by the particular bile salts occurring in their hosts. Tkachuck and MacInnis (1971) and Surgan and Roberts (1976) have shown that some purified bile salts directly influence the metabolic activities of the tapeworms *Hymenolepis diminuta* and *H. microstoma*; they may do this by affecting the surface of the parasites.

In unpublished work with me, R. J. Jørgensen found that purified bile salts from ox bile would affect the mobility of the infective stage of the cattle lungworm *Dictyocaulus viviparus*. Since *Dictyocaulus* larvae do not respond when exposed to other digestive juices or to changes in pH and temperature, it is most likely that bile salts are essential in the activation of ingested larvae and their subsequent penetration through the gut wall of the host animal; see also Jørgensen (1973).

Bile salts may affect the growth and behaviour of intestinal microorganisms (e.g. Fernandes and Smith, 1977). In recent years, M. J. Hill and his colleagues have re-investigated the long-held idea that certain intestinal organisms might be capable of aromatizing some of the six-membered rings in bile acids and steroid hormones, so producing compounds that could induce or facilitate the production of bowel and breast cancer, as the highly carcinogenic methylcholanthrene chemically derived from deoxycholic acid would undoubtedly do (Hill, 1976).

Nigro and Campbell (1976) have reviewed their studies and those of others on effect of diet on production of intestinal tumours by carcinogens and conclude that dietary effects might involve bile salt secretion and degradation by gut microorganisms.

1.11. Bile salts in tissues other than liver

There is evidence that lithocholic acid (p. 56) and possibly other bile acids can occur in the brains of patients with multiple sclerosis and of animals with induced demyelinating disease (Nicholas, 1976). Dupont et al. (1976) found mass spectral peaks corresponding to those of bile acids in extracts from skeletal muscle, adipose tissue, kidney, pancreas and brain, all obtained from rats. These workers concluded that the above-mentioned tissues together might contain a total of about 3 mg of bile acids. Bile acids were also identified by enzyme-fluorimetry (p. 126). It is not suggested that any tissue other than liver can biosynthesise bile salts and they might all have a hepatic origin.

References

Blumenthal, S. G., Taggart, D. B., Ikeda, R. M., Ruebner, B. H. and Bergstrom, D. E. (1977) Conjugated and unconjugated bilirubins in bile of humans and rhesus monkeys. Structure of adult human and rhesus-monkey bilirubins compared with dog bilirubins. Biochem. J., 167, 535—548.

Borgström, B. (1977) The action of bile salts and other detergents on pancreatic lipase and the interaction with colipase. Biochim. Biophys. Acta, 488, 381—391.

Bouchier, I. A. D. and Cooperband, S. R. (1967) Sephadex filtration of a macromolecular aggregate associated with bilirubin. Clin. Chim. Acta, 15, 303—313.

Clausen, J., Kruse, I. and Dam, H. (1965) Fractionation and characterisation of proteins and lipids in bile. Scand. J. Clin. Invest., 17, 325—335.

Desmet, V. (1976) Spoken at Falk Symposium No. 23, Basle.

Diamond, J. M. (1968) Transport mechanisms in the gallbladder. In: Handbook of Physiology. Section 6; Alimentary Canal. Ed. C. F. Code. Vol. V., pp. 2451—2482. American Physiological Society, Washington D.C.

Dietschy, J. M. (1968) Mechanisms for the intestinal absorbtion of bile acids. J. Lipid Res., 9, 297—309.

Dowling, R. H. (1972) The enterohepatic circulation. Gastroenterology, 62, 122—140.

Dupont, J., Suk Yon Oh and Janson, P. (1976) Tissue distribution of bile acids: methodology and quantification. In: The Bile Acids. Eds. P. P. Nair and D. Kritchevsky. Vol. 3, pp. 17—28. Plenum Press, London—New York.

Erhlinger, S. and Dhumeaux, M. D. (1974) Mechanisms and control of secretion of bile water and electrolytes. Gastroenterology, 66, 281—304.

Espinsoa, E. (1976) Circulating tissue antigens. III. Identification and characterisation of antigens of limited and of wide body distribution in human gall bladder bile. Presence of serum of patients with acute hepatitis. Clin. Exp. Immunol., 25, 410—417.

Etter-Kjesaas, H. and Kuenzle, C. C. (1975) A polypeptide conjugate of bilirubin from human bile. Biochim. Biophys. Acta, 400, 83—94.

Fernandes, P. B. and Smith, Jr., H. J. (1977) The effect of anaerobiosis and bile salts on the growth and toxin production by *Vibrio cholerae*. J. Gen. Microbiol., 98, 77—86.

Fevery, J., Van de Vijver, M., Michiels, R. and Heirwegh, K. P. M. (1977) Comparison in different species of biliary bilirubin-IX α conjugates with the activities of hepatic and renal bilirubin-IXα-uridine diphosphate glycosyltransferases. Biochem. J., 164, 737—746.

Gent, W. L. G., Haslewood, G. A. D. and Montesdeoca, G. (1971) Macromolecular compounds of bilirubin in human bile. Biochem. J., 122, 15—16P.

Gordon, E. R., Goresky, C. A., Chang, T-H. and Perlin, A. S. (1976) The isolation and characterisation of bilirubin diglucuronide, the major bilirubin conjugate in dog and human bile. Biochem. J., 155, 477—486.

Harrison, F. A. and Leat, W. M. F. (1975) Digestion and absorption of lipids in non-ruminant and ruminant animals: a comparison. Proc. Nutr. Soc., 34, 203—210.

Heirwegh, K. P. M., Fevery, J., Michiels, R., Van Hees, G. P. and Compernolle, F. (1975) Separation by thin-layer chromatography and structure elucidation of bilirubin conjugates isolated from dog bile. Biochem. J., 145, 185—199.

Hill, M. J. (1976) Fecal steroids in the etiology of large bowel cancer. In: The Bile Acids. Eds. P. P. Nair and D. Kritchevsky. Vol. 3, pp. 169—200. Plenum Press, New York—London.

Jansen, F. H. and Billing, B. H. (1971) The identification of monoconjugates of bilirubin in bile as amide derivatives. Biochem. J., 125, 917—919.

Javitt, N. B., Rifkind, A. and Kappas, K. (1973) Porphyrin-heme pathway: regulation by intermediates in bile acid synthesis. Science, 182, 841—842.

Jørgensen, R. J. (1973) In vitro effect of bile on the motility of *Dictyocaulus viviparus* third stage larvae. Acta Vet. Scand., 14, 341—343.

Kuenzle, C. C. (1970) Bilirubin conjugates of human bile etc. Biochem. J., 119, 387—435.

Leat, W. M. and Harrison, F. A. (1977) The relative importance of bile salts and phospholipids in fat absorption in the sheep. Proc. Nutr. Soc., 36, 70A.

Lough, A. K. and Smith, A. (1976) Influence of the products of phospholipolysis of phosphatidylcholine on micellular solubilisation of fatty acids in the presence of bile salts. Br. J. Nutr., 35, 77—96.

Marks, J. W., Bonorris, G. G. and Schoenfield, L. J. (1976) Pathophysiology and dissolution of cholesterol gallstones. In: The Bile Acids. Eds. P. P. Nair and D. Kritchevsky. Vol. 3, pp. 81—113. Plenum Press, New York—London.

Nakayama, F. and van der Linden, W. (1970) Bile from gallbladder harbouring gallstone: can it indicate stone formation? Acta Chir. Scand., 136, 605—610.

Nair, P. P. (1976) Bile-salt protein interaction. In: The Bile Acids. Eds. P. P. Nair and D. Kritchevsky. Vol. 3, pp. 29—52. Plenum Press, New York—London.

Nicholas, H. J. (1976) Bile acids and brain. In: The Bile Acids. Eds. P. P. Nair and D. Kritchevsky. Vol. 3, pp. 1—15. Plenum Press, New York—London.

Nigro, N. D. and Campbell, R. L. (1976) Bile acids and intestinal cancer. In: The Bile Acids. Eds. P. P. Nair and D. Kritchevsky. Vol. 3, pp. 155—168. Plenum Press, New York—London.

Normann, Per. T. and Norum, K. R. (1976) Newly synthesized hepatic cholesterol as precursor for cholesterol and bile acids. Scand. J. Gastroenterol., 11, 427—432.

Percy-Robb, I. W. (1975) Bile acid synthesis: an alternative pathway leading to hepatotoxic compounds? Essays Med. Biochem., 1, 59—80.

Pope, J. L., Parkinson, T. M. and Olsen, J. A. (1966) Action of bile salts on the metabolism and transport of water-soluble nutrients by perfused rat jejunum in vitro. Biochim. Biophys. Acta, 130, 218—232.

Porter, K. R. (1969) Independence of fat absorption and pinocytosis. Fed. Proc. 28, 35—40.

Rappaport, A. M. (1975) The microcirculatory hepatic unit. In: Drugs and the Liver. Eds. W. Gerok and K. Sickinger. pp. 425—434. F. K. Schattauer Verlag, Stuttgart—New York.

Reichen, J., Preisig, R. and Paumgartner, G. (1977) Influence of chemical structure on hepatocellular uptake of bile acids. In: Bile Acid Metabolism in Health and Disease. Eds. G. Paumgartner and A. Stiehl. pp. 113—123. MTP Press, Lancaster.

Rose, R. C., Gerlarden, R. T. and Nahrwold, D. L. (1973) Electrical properties of isolated human gallbladder. Am. J. Physiol. 224, 1320—1326.

Senior, J. R. (1964) Intestinal absorption of fats. J. Lipid. Res., 3, 495—521.

Schaffner, F. (1975) The ultrastructure of bile secretion. In: Drugs and the Liver. Eds. W. Gerok and K. Sickinger. pp. 91—100. F. K. Schattauer Verlag, Stuttgart—New York.

Schrager, J. (1970) The chemical composition and function of gastrointestinal mucus. Gut, 11, 450—456.

Simon, F. R., Sutherland, E., Accatino, L., Vial, J. and Mills, D. (1977) Studies on drug-induced cholestasis: effect of ethinyl estradiol on hepatic bile acid receptors and (Na^+-K^+)-ATPase. In: Bile Acid Metabolism in Health and Disease. Eds. G. Paumgartner and A. Stiehl. pp. 133—143. MTP Press, Lancaster.

Smith, R. L. (1973) The Excretory Function of Bile. Chapman and Hall, London.

Smyth, J. D. and Haslewood, G. A. D. (1963) The biochemistry of bile as a factor in determining host specificity in intestinal parasites, with particular reference to *Echinococcus granulosus*. Ann. N.Y. Acad. Sci., 113, 234—260.

Sobotka, H. (1937) Physiological Chemistry of the Bile. pp. 1—11. Baillière, Tindall and Cox, London.

Surgan, M. H. and Roberts, L. S. (1976) Adsorption of bile salts by the cestodes *Hymenolepis diminuta* and *H. microstoma* etc. J. Parasitol., 62, 78—86 and 87—93.

Tanaka, N., Portman, O. W. and Osuga, T. (1976) Effect of type of dietary fat, cholesterol and chenodeoxycholic acid on gallstone formation, bile acid kinetics and plasma lipids in squirrel monkeys. J. Nutr., 106, 1123—1134.

Tkachuck, R. D. and MacInnis, A. J. (1971) The effect of bile salts on the carbohydrate metabolism of two species of hymenolepidid cestodes. Comp. Biochem. Physiol., 40B, 993—1003.

CHAPTER 2

Chemistry and methods of separation and analysis

Chemistry

2.1. Definitions

Bile salts are here defined as substances secreted by the liver into the bile and presumably aiding in the digestion of fats and their absorption from the intestine more or less as described in Chapter 1. This definition is taken to include numerous minor biliary constituents present in such small amounts as to make their function appear doubtful, but chemically so obviously closely related to the chief bile salts as to leave no doubt that they arise by the same or very similar biochemical processes. In animals having a liver and bile (all vertebrates) the bile salts are steroids. Physiologically, they may be *primary*, i.e. made in the liver (presumably from cholesterol or some other sterol in all cases) or *secondary*, a term used to mean primary bile salts modified in some way by microorganisms during the entero-hepatic circulation (Fig. 1.1) and perhaps further altered on their return to the liver (Chapter 3).

2.2. General nature of bile salts

Steroid bile salts have the characteristic nucleus and all or part of the side-chain shown in *cholestane* (Formula 2.1).

In this structure, the methyl group at C-10 is conventionally shown as β-orientated (projecting above the plane of the paper); the methyl

2.1 Cholestane.

group at C-13 and the side-chain at C-17 have the same orientation. The hydrogen at C-5 may be β or α (shown by a broken line), meaning that the A/B ring junction is respectively *cis* or *trans*; the B/C and C/D ring junctions are *trans* so that the hydrogens at C-9 and C-14 are α, as shown. In the numbering shown in Formula 2.1, C-26 is defined as the CH_3 group present as such throughout the biosynthesis of cholesterol from mevalonic acid. The configuration at C-20 is R. The hydrocarbon corresponding to structure 2.1, up to and including C-24, is called *cholane*.

In all animals examined, the pH of the bile is such that the bile salts in it will be present entirely or almost entirely as anions; the counterions are Na^+, K^+ or, in minor amounts, Ca^{2+} and perhaps Mg^{2+}. Thus, isolation will give usually sodium or potassium salts. The steroid anions are those of the following chemical types:

    A. Sulphate esters of bile alchohols
    B. Taurine conjugates of C_{27} bile acids
    C. Taurine conjugates of C_{24} bile acids
    D. Glycine conjugates of C_{24} bile acids
    E. The anions of unconjugated C_{28} (in one case), C_{27} and C_{24} bile acids.

The distribution of these types in various animal forms is considered in detail later.

2.3. A. Bile alcohol sulphates

The trivial names of the alcohols include as a prefix part of the Latin name of the genus of the animal from which they were first isolated.

This practice follows the example of Olof Hammarsten who, in 1898, isolated an alcohol which he called "scymnol" after alkaline hydrolysis of the bile of a shark, *Scymnus borealis*. The chemistry of Hammarsten's "α-scymnol" remained obscure until 1962, when the parent hexol was made from a sample of Hammarsten's original sulphate and also partially synthesised from cholic acid (Formula 2.4). The earlier history of scymnol has already been given (Matschiner, 1971). Scymnol sulphate (Formula 2.2) is the 26 (or 27) sulphate of 5β-cholestane-3α,7α,12α,24ξ,26,27-hexol.

2.2 Scymnol sulphate [(5β-Cholestane-3α,7α,12α,24ξ,26,27-hexol-26 (or 27)-sulphate)*.

In many cases both 5α and 5β epimers of the alcohols occur as sulphates in the bile; the prefix 5α or 5β is then used with the original name.

A list of natural bile alcohols known at the present time, together with the location in the molecule of the sulphate ester groups, where determined, is given in Table 2.1. Hoshita and Kazuno's (1968) and Matschiner's (1971) reviews give further details of the history and properties of substances isolated up to these dates. All bile alcohols detected seem to be of primary origin, as judged by their structure and by the known reactions brought about by intestinal microorganisms.

It will be seen that some of the substances listed in Table 2.1 have not actually been isolated as such; their presence was deduced from the

---

*The sign ξ means that configuration is unknown or uncertain and is shown by a wavy line, as at C-24 in the formula.

TABLE 2.1
Bile alcohols and their natural sulphates

Trivial name of alcohol	M.p. (°C)	$[\alpha]_D$	Systematic name (Text Formula)	Location of sulphate	Natural Source	References (when not in Matschiner, 1971)
Arapaimol-A	220	+34	5β, 25ξ-Cholestane-2β, 3α,7α,12α,26-pentol (4.7, p. 92)	Unknown	*Arapaima gigas*; some frogs	Haslewood and Tökés, 1972; Anderson et al., 1974
Arapaimol-B	230	+30	5β-Cholestane-2β,3α,7α,12α, 26,27-hexol	Unknown	*Arapaima gigas*	Haslewood and Tökés, 1972
5α-Bufol	241	+38	5α,25ξ-Cholestane-3α,7α,12α, 25,26-pentol (4.9, p. 94)	Probably C-26	Lungfish; a newt; several frogs	Anderson et al., 1974; Amos et al., 1977; Anderson et al., 1974
5β-Bufol	178	+38	5β,25ξ-Cholestane-3α,7α,12α, 25,26-pentol	Probably C-26	Toads of genus *Bufo*; some frogs	Anderson et al., 1974
5α-Chimaerol	235	+33.5	5α,25ξ-Cholestane-3α,7α,12α, 24(+),26-pentol (sulphate, 4.6, p. 90)	C-26	Fishes of families Catostomidae and Cyprinidae; lungfish	Anderson and Haslewood, 1970; Amos et al., 1977; unpublished observations
5β-Chimaerol	182	+42	5β,25ξ-Cholestane-3α,7α,12α, 24(+),26-pentol (4.2, p. 82)	Unknown	*Chimaera monstrosa*; some sharks and rays	Anderson and Haslewood, 1970
5α-Cyprinol	244	+29	5α-Cholestane-3α,7α,12α, 26,27-pentol	C-26 (or 27)	Fishes of order Ostariophysi; *Latimeria chalumnae*; lungfish; a salamander; some frogs	Tammar (Review) 1974a; Anderson et al., 1974; Amos et al., 1977
5β-Cyprinol	173	+36	5β-Cholestane-3α,7α,12α, 26,27-pentol (4.4, p. 87)	Probably C-26 (or 87)	Several bony fishes; some frogs	Tammar. 1974a; Anderson et al., 1974
	189	+37	5β-Cholestane-3α,7α,12α, 25-tetrol	None		Setogouchi et al., 1974
	211	+49	5β-Cholestane-3α,7α,12α, 23S,25-pentol	None	Bile and faeces from cerebrotendinous xanthomatosis patients	Hoshita et al., 1976
	214	+45	5β-Cholestane-3α,7α,12α, 24R,25-pentol	None		Shefer et al, 1975
27-Deoxy-5α-cyprinol (= 24-deoxy-5α-chimaerol = 25-deoxy-5α-	232	+39	5α,25ξ-Cholestane-3α,7α, 12α,26-tetrol (4.5, p. 89)	Unknown	Lungfish; carp and probably other ostariophysan fishes; some frogs and toads	Amos et al., 1977; Anderson et al., 1974

TABLE 2.1 (continued)

Trivial name of alcohol	M.p. (°C)	[α] D	Systemic name (Text Formula)	Location of sulphate	Natural Source	References (when not in Matschiner, 1971)
27-Deoxy-5β-cyprinol			5β,25ξ-Cholestane-3α,7α,12α,26-tetrol	Unknown	Some frogs and toads	Anderson et al., 1974
5α-Dermophol			5α-Cholestane-3α,7α,12α,25,26,27-hexol	Unknown	Some amphibians	Kihara et al., 1977
5β-Dermophol			5β-Cholestane-3α,7α,12α,25,26,27-hexol (4.8, p. 93)	Unknown		
Latimerol	236	+33	5α-Cholestane-3β,7α,12α,26,27-pentol(sulphate, 4.3, p. 84)	Probably C-26 (or 27)	Coelacanth, *Latimeria chalumnae*	Amos et al., 1977
Myxinol	204	−15	5α,25ξ-Cholestane-3β,7α,16α,26-tetrol (2.8, p. 40)	C-3 and C-26 (or 27)	Hagfishes (Myxinidae)	
16-Deoxymyxinol	219	+13	5α,25ξ-Cholestane-3β,7α,26-triol	Unknown	Hagfishes	
5α-Petromyzonal	231 (241)	+27.5	5α-Cholane-3α,7α,12α,24-tetrol (4.1, p. 81)	C-24	Lampreys; lungfish	Amos et al., 1977
5β-Petromyzonal	227		5β-Cholane-3α,7α,12α,24-tetrol	Unknown	Lungfish; amphibians	Amos et al., 1977
5α-Ranol	Amorphous	+21	27-Nor-5α-cholestane-3α,7α,12α,24ξ,26-pentol (sulphate 4.10, p. 95)	C-24	Several frogs; *Salamandra salamandra*	Anderson et al., 1974; unpublished observations
5β-Ranol	189		27-Nor-5β-cholestane-3α,7α,12α,24ξ,26-pentol	C-24	Several frogs	Anderson et al., 1974
26-Deoxy-5α-ranols	Amorphous		27-Nor-5α-cholestane-3α,7α,12α,24α & 24β-tetrols	Unknown	Several frogs	Anderson et al., 1974 Noma et al., 1976
26-Deoxy-5β-ranols	(24α) 195 (24β) 182	+35 +27	27-Nor-5β-cholestane-3α,7α,12α,24α & 24β-tetrols	Unknown	Several frogs	Anderson et al., 1974 Noma et al., 1976
26-Deoxy-26-nor-5α-ranol			26,27-Bisnor-5α-cholestane-3α,7α,12α,24ξ-tetrol	Unknown	Some frogs	Anderson et al., 1974
Scymnol	123 (hydrate)	+34	5β-Cholestane-3α,7α,12α,24ξ,26,27-hexol (sulphate, 2.2)	C-26 (or 27)	Sharks and rays	Anderson et al., 1974

mass spectra (MS) of fractions eluted during gas-liquid chromatography (GLC). Unacceptable as this sort of evidence might have been to organic chemists of the old school, who insisted on adequate elemental analyses and characterisation by physical properties of compounds claimed as new, GLC—MS results alone are now taken as proof for the existence and structure of organic compounds. The GLC—MS technique is capable of discovering and identifying even quite small traces of compounds in mixtures and its application to bile alcohol and bile acid mixtures has greatly increased our understanding of what the enzymes catalysing the production of these substances can do in various biological circumstances (Anderson et al., 1974; Danielsson and Sjövall, 1975). At the time of writing GLC—MS is not, unfortunately, applicable to bile salts themselves since these are involatile at temperatures not causing complete decomposition. However, high-pressure liquid chromatography (e.g. Shaw and Elliott, 1976) combined with mass spectroscopy can be expected eventually to extend the advantages of GLC—MS to bile salts and this will be a substantial advance, especially for taurine-conjugated C_{27} bile acids which require conditions for chemical hydrolysis so drastic as to cause considerable changes in the molecule. Solvolysis of bile alcohol sulphates, on the other hand, can be effected by very mild chemical methods not believed to alter the bile alcohol.

Three bile alcohols, 5β-cholestane-3α,7α,12α,25-tetrol, 5β-cholestane-3α,7α,12α,23S,25-pentol and 5β-cholestane-3α,7α,12α,24R,25-pentol found in considerable amounts in human bile and faeces in the disease cerebrotendinous xanthomatosis (Shefer et al., 1975), are more fully discussed in Chapters 3 and 5.

2.4. B. Taurine conjugates of C_{27} acids

This type of bile salt is common in some groups of reptiles and amphibians; it is also found in a few fishes. The acids themselves possess the carbon skeleton of cholesterol and can be regarded as derived from that substance by (a) conversion of the ring nucleus to that of chenodeoxycholic acid, cholic acid or its 5α epimer, (b) oxidation of C-26 to COOH and (c) further hydroxylation, in some cases, of the ring nucleus or side-chain (see Chapter 3).

Conjugates with taurine only have been found. The only published evidence for secondary C_{27} acids is the presence in Green turtle bile of a

TABLE 2.2
C_{27} Bile acids found as taurine conjugates

Trivial name or empirical formula of acid	Systematic full or partial structure (Text Formula)	Origin	References (when not in Matschiner, 1971)
Arapaimic	$2\beta,3\alpha,7\alpha,12\alpha$-Tetrahydroxy-$5\beta$-cholestan-26 (or 27)-oic acid	*Arapaima gigas*	Haslewood and Tökés, 1972
Tetrahydroxysterocholanic and tetrahydroxyisosterocholanic	The existence of two isomeric acids is doubtful: one is certainly $3\alpha,7\alpha,12\alpha,22\xi$-tetrahydroxy-$5\beta$-cholestan-26 (or 27)-oic acid (4.11, p. 97)	Turtles and tortoises	Haslewood et al., 1978
Varanic	$3\alpha,7\alpha,12\alpha,24\xi$-Tetrahydroxy-$5\beta$-cholestan-26(or 27)-oic acid (4.12, p. 98)	Lizards of families Varanidae and Helodermatidae	Haslewood, 1967
2 acids, $C_{27}H_{46}O_6$	Both form lactones and have the cholic acid ring nucleus	Eel, *Anguilla japonica*	Okada et al., 1962
	$3\alpha,7\alpha,12\alpha$-Trihydroxy-5α-cholestan-26(or 27)-oic acid	Iguana; mud-puppy	Okuda et al., 1972; Ali, 1975
Trihydroxycoprostanic	$3\alpha,7\alpha,12\alpha$-Trihydroxy-5β-cholan-26-oic acid (The 27-oic acid may occur but isolated material may be the result of racemization during alkaline hydrolysis) (3.9, p. 50)	Some amphibians; reptiles, birds and mammals; man (especially in biliary atresia)	See Chapters 3, 4 and 5
	$3\alpha,7\alpha$-Dihydroxy-5β-cholestan-26(or 27)-oic acid	*Alligator mississippiensis*; man	Hanson and Williams, 1971
	$3\alpha,12\alpha,22\xi$-Trihydroxy-5β-cholestan-26 (or 27)-oic acid	Green turtle, *Chelonia mydas*	Haslewood et al., 1978

taurodeoxycholate analogue of the principal (presumably) primary C_{27} acid (Haslewood et al., 1978). A list of C_{27} bile acids found as taurine conjugates is given in Table 2.2.

2.5. C and D. Taurine and glycine conjugates of C_{24} bile acids

The C_{24} bile acids are the longest known as well as the commonest in the animal kingdom. Some scientists and others still appear to regard these acids as the only "bile acids" and do not seem to recognise the existence (and predominance in certain animal forms) of other bile salt types. This is particularly true in a few medical quarters where, of course, the chief interest is in the bile of man and the common laboratory animals which have no more than traces of any but C_{24} acids. As discussed later, the C_{24} acids can be regarded as the end-points (so far) of bile acid evolution and are common to most advanced animal forms, with the conspicuous exception of elasmobranch fishes and amphibians. Thus they are characteristic of most teleostean fishes, most advanced lizards, snakes, birds and mammals.

To Lindstedt and Sjövall (1957) belongs the credit of demonstrating the existence of secondary C_{24} bile acids; they found that when a biliary fistula was maintained in a rabbit, the glycine-conjugated deoxycholic acid which is normally the chief bile salt of gall-bladder bile in this species was gradually entirely replaced by glycine-conjugated cholic acid (Formula 2.3).

2.3 Cholic acid ($3\alpha,7\alpha,12\alpha$-trihydroxy-5β-cholan-24-oic acid).

Lindstedt and Sjövall considered the possibility that the deoxycholic acid normally present had been replaced by cholic acid because of the changes in physiology caused by drainage of bile through the fistula, but concluded that it was more likely that intestinal microorganisms usually removed the 7α-OH group from cholic acid to give deoxycholic acid in the intact animal. All later work has confirmed this conclusion and there is no evidence that deoxycholic acid (Formula 2.4) can ever be a primary bile acid. The effect of the enterohepatic circulation is further discussed in Chapter 3.

2.4 Deoxycholic acid |(3α,12α-dihydroxy-5β-cholan-24-oic acid).

Table 2.3 gives a list of C_{24} acids conjugated in bile with taurine or glycine to give bile salts. Glycine conjugation has been found only in eutherian (placental) mammals. In Table 2.3, the stem (unsubstituted) acid is given as cholan-24-oic acid, which is the name recommended by the International Union of Pure and Applied Chemistry (IUPAC) and International Union of Biochemistry (IUB). The IUB—IUPAC Rules* also state that the configuration (α or β) at C-5 must be given in systematic names. Heinrich Wieland first made the stem-acid of the common bile acids and called it "cholanic acid"; this name with the 5α or 5β prefix is a suitable basis for trivial names of bile acids not christened by their discoverers. The term "cholanoic acid" for this purpose has no sanction in history or in the IUB—IUPAC rules and it will not be used in this book. Following the original German

---

*Biochem. J., 113, 5-28 (1969).

TABLE 2.3
Cholan-24-oic acids found conjugated with taurine or glycine to give bile salts

Trivial name of acid (Text Formula)	Configuration at C-5	Position and configuration of OH groups	Approximate melting points °C	$[\alpha]_D^0$	Source	Remarks and References (when not in Matschiner, 1971)
$3\alpha,7\alpha,12\alpha,23\xi$-Tetrahydroxy-$5\beta$-cholanic (4.14, p. 100)	β		Amorphous	+36	Pinnipedia; some snakes	Ikawa and Tammar, 1976
Allocholic	α	$3\alpha,7\alpha,12\alpha$	240	+23	Some fishes; some amphibians; many lizards; birds; mammals	Elliott, 1971; Eyssen et al., 1976
Bitocholic (4.14, p. 100)	β	$3\alpha,12\alpha,23\xi$	Amorphous	+48	Some snakes	Secondary: Ikawa and Tammar, 1976
Cholic (2.3, p. 28)	β	$3\alpha,7\alpha,12\alpha$	198	+37	The commonest bile acid	
Haemulcholic	β	$3\alpha,7\alpha,22\alpha_F$	252	−10	Teleost, *Parapristipoma trilineatun*; other fishes	Hoshita et al., 1967; Appendix
Hyocholic (4.19, p. 114)	β	$3\alpha,6\alpha,7\alpha$	189	+5.5	Pigs (*Sus*)	Hsia, 1971
α-Muricholic (4.16, p. 108)	β	$3\alpha,6\beta,7\alpha$	200	+38	Laboratory rats and mice	Hsia, 1971; Eyssen et al., 1976
β-Muricholic (4.16, p. 108)	β	$3\alpha,6\beta,7\beta$	228	+62	Laboratory rats and mice	Hsia, 1971; Eyssen et al., 1976
ω-Muricholic	β	$3\alpha,6\alpha,7\beta$	191	+36	Laboratory rats and mice	Secondary: Hsia, 1971; Eyssen et al., 1976
Phocaecholic (4.18, p. 112)	β	$3\alpha,7\alpha,23\xi$	225	+11	Pinnipedia	
Pythocholic (4.13, p. 99)	β	$3\alpha,12\alpha,16\alpha$	187	+28 (methyl ester)	Boid snakes	Secondary
Chenodeoxycholic (3.16, p. 56)	β	$3\alpha,7\alpha$	about 140 (solvated)	+11.5	Almost as common as cholic acid	Of great medical interest (Chapter 5)
Deoxycholic (2.4, p. 29)	β	$3\alpha,12\alpha$	177 usually solvated	+53	Many, if not most, mammals; some lower forms	Secondary

TABLE 2.3 (continued)

Trivial name of acid (Text Formula)	Configuration at C-5	Position and configuration of OH groups	Approximate melting points °C	$[\alpha]_D^0$	Source	Remarks and References (when not in Matschiner, 1971)
Hyodeoxycholic (4.20, p. 114)	β	3α,6α	197	+5	Pigs (*Sus*); wart hog	Chiefly secondary, but partly primary in pigs (Haslewood, 1971)
Lagodeoxycholic	α	3α,12α	218	+37	Laboratory rabbit	Secondary: Elliott, 1971
Ursodeoxycholic (4.17, p. 111)	β	3α,7β	203	+57	Several mammals	Probably usually secondary; of great medical interest
3α,6β-Dihydroxy-5β-cholanic	β		210	+37	Pigs (*Sus*)	Probably secondary
3β,6α-Dihydroxy-5β-cholanic	β		190	+5	Pigs (*Sus*)	Probably secondary
3α-Hydroxy-6-oxo-5α-cholanic	α		194	−9	Pigs (*Sus*)	Secondary and probably in bile as the 5β-epimer
3α-Hydroxy-7-oxo-5β-cholanic: 7-ketolithocholic (taurine conjugate, 4.15, p. 105)	β		203	−27	Many mammals; some birds	Probably mainly secondary
Lithocholic (3.15, p. 56)	β	3α	186	+35	Many mammals and some birds, in minor amounts	Probably almost always secondary: of great medical interest (Chapter 5)

precedent, "cholic" acids contain three hydroxyl groups in the molecule and thus, "deoxycholic" acids two such groups.

As well as the substances listed in Table 2.3, several other cholanic acids have been isolated from bile, urine and faeces. These are all believed to be the result of disease or of microbial action. The mixture in faeces is, not surprisingly, extremely complex and variable; it is doubtful whether the police investigator who once asked me if it would be possible to identify a criminal by the pattern of his faecal bile acids had hit upon a useful notion. Some of the C_{24} acids found in human bile and excreta are mentioned in Chapter 5: others found in various species are listed by Van Belle (1965), Haslewood (1967) and Tammar (1974a,b,c). Reviews of the changes caused by microorganisms have been given by Hayakawa (1973) and Kellogg (1973). Van Belle (1965) also compiled very useful tables of the properties and derivatives of cholanic acids made in the laboratory as well as those discovered in nature.

2.6. E. Unconjugated bile acids acting as bile salts

The presence of unconjugated bile acids in bile at once arouses the suspicion that they may be the result of disease or of post-mortem bacterial changes. This is particularly likely for the 5β-cholanic acids, for many microorganisms are known that will remove the taurine or glycine from their conjugates (for example, see Aries and Hill, 1970a; Kellogg, 1973) and an enzyme from *Clostridium perfringens* that will hydrolyse glycine and taurine conjugates of cholic, chenodeoxycholic, deoxycholic and other 5β-cholanic acids is commercially available (Nair, 1973).

Nevertheless, there are unconjugated bile acid ions in some biles, collected in such a way as to prevent post-mortem changes and from apparently healthy animals. For example, trihydroxycoprostanate (p. 50) is a feature of some frog biles (Kuramoto et al., 1973; Anderson et al., 1974). The taurine conjugate of this acid and of its C-24 hydroxyl derivative, varanic acid or its isomers (also found in a frog in the unconjugated condition by Kuramoto et al., 1973), cannot be hydrolysed by the bacterial enzymes that split conjugates of 5β cholanic acids. Indeed, B. S. Drasar and I, after a long search, found only a few microorganisms that would hydrolyse taurine-conjugated C_{27} acids, so that it seems rather unlikely that these unconjugated acids could arise in the bile by microbial action, especially since no other evidence of an enterohepatic circulation of amphibian bile salts is detectable in their chemistry. It seems more probable that the trihydroxycoprostanate ion is used as a bile salt prior to the development of an effective conjugating system: clearly this idea could be experimentally tested.

Of five species of toads of the genus *Bufo* examined, only *Bufo b. formosus* was found to have unconjugated trihydroxybufosterocholenic acid (3α,7α,12α-trihydroxy-5β-cholest-22-ene-24-carboxylic acid, Formula 2.5) in the bile (Anderson et al., 1974). The structure of this acid was, after prolonged investigation, finally elucidated by Hoshita et al. (1967) and confirmed by Anderson et al. (1974). Toads injected with ^{14}C-4-cholesterol or ^{14}C-2-mevalonate did not form radioactive trihydroxybufosterocholenic acid (Kuramoto et al., 1974) and its biochemical origin remains a mystery. It may of course arise from some C_{28} sterol peculiar to the diet of this Japanese toad and might be unacceptable to a biochemical system for making conventional taurine conjugates.

2.5 Trihydroxybufosterocholenic acid (3α,7α,12α-trihydroxy-5β-cholest-22-ene-24-carboxylic acid).

This toad also contains unconjugated 3α,7α,12α-trihydroxy-5β-cholest-23-en-26 (or 27)-oic acid (Kuramoto et al., 1974).

I found a sample of South American piranha (probably *Serrasalmus ternetzi*) bile to contain only cholate as its bile salt, but this observation needs confirmation on a sample freshly collected in conditions excluding post-mortem changes. Confirmation would show that C_{24} bile acid ions themselves can function effectively as bile salts. Peric-Golia and Socic (1968) found some free cholic acid in bile of healthy sheep and similar findings have been made in healthy men. My colleagues and I have also found high proportions of C_{24} bile acid ions in fishes of the genus *Synodontis*.

In summary, several cases are known amongst amphibians in which the ions of C_{27} acids listed in Table 2.2 are present in bile in proportions such that they must be acting physiologically as bile salts. In one case, the anions of a C_{28} acid of unknown biochemical origin are present. An unconfirmed finding suggests that cholate itself may be capable of acting in vivo as a bile salt.

2.7. Other conjugates

Peric-Golia and Jones (1962) showed that conjugates of 5β-cholanic acids with ornithine could be formed by guinea pigs and rats, especially after injection with a polysaccharide from *Klebsiella pneumoniae*; such conjugates had been found in the bile of patients infected with this organism. Gordon et al. (1963) found similar conjugates in ox and human bile. Yousef and Fisher (1975) and Myher et al. (1975) have synthesized arginine, histidine and ornithine conjugates of cholic acid and shown that the arginine conjugate is the one secreted in bile by isolated perfused rat livers.

Palmer (1967) was the first to show that formation of sulphate esters, in this case of lithocholic acid and its conjugates, was a metabolic pathway for C_{24} bile acids. It has since become clear that variable amounts of sulphate esters of the common cholic, chenodeoxycholic and deoxycholic acids and their conjugates are to be found in bile, intestine and urine, especially in diseases of the liver and biliary tree.

Parmentier and Eyssen (1975) made the 3-, 7- and 12-monosulphates of cholic acid and Eyssen et al. (1976) found that in both germ-free and conventional mice, cholic acid and chenodeoxycholic acids were present in the bile and faeces partially as their 7-sulphates; other bile acids were also sulphated. Summerfield et al. (1976) prepared the 7-mono- and 3,7-disulphates of chenodeoxycholic acid and showed that an isolated perfused rat kidney could make the former substance, as well as traces of chenodeoxycholic acid 3-sulphate. Haslewood and Haslewood (1976a) made the 3-sulphates of cholic, chenodeoxycholic and deoxycholic acids as crystalline disodium salts (e.g. Formula 2.6).

Tamari et al. (1976) found "ciliatocholate", a phosphonic acid analogue (R.CO.NH.CH$_2$.CH$_2$.PO$_3^{2-}$: R.CO = cholyl) of taurocholate as a minor constituent of ox bile salts.

2.6 Cholic acid 3-sulphate disodium salt.

The purity of some of the preparations described is questionable and much more chemical work is required in this field.

Summerfield et al. (1976) are amongst those who have described methods for isolating bile salts from urine and applying techniques for fractionation of bile acid conjugates and sulphates on Sephadex LH-20. An excellent solvent system for separating sulphates on thin-layer chromatograms is that of Cass et al. (1975).

Back and Bowen (1976) have made 3-β-D-monoglucuronides of cholic, chenodeoxycholic, deoxycholic and lithocholic acid. Back and his colleagues (e.g. Back, 1976) describe the isolation of fractions containing bile acid glucuronides from urine and blood plasma.

The occurrence of sulphates and glucuronides in human bile and urine is further discussed in Chapter 5.

2.8. Methods of separation and analysis

Bile salts may be preserved by the addition of the bile (or gall-bladders containing it) to at least 4 volumes of ethanol or ethanol-methanol (methylated spirits) as soon as possible after death or removal from the body. In this way, I have found bile salts apparently to keep indefinitely, although aqueous solutions soon grow microorganisms which affect them. The excess of alcohol precipitates mucin and when the bile salts are required addition of further ethanol and filtration will remove, on the filter, the mucin and some of the bile pigment. The filtrate is evaporated to dryness and lipid may now be almost completely extracted without loss of bile salts by two or three washings with light petroleum (b.p. 40—60°C). The petroleum-insoluble material is extracted with methanol, in which bile salts of every type readily dissolve, and the filtered methanol extract is evaporated to dryness. The residue of crude bile salts may be a more or less crystalline solid, a foam (after drying in vacuo) or a gum. In many cases this mixture is hygroscopic, sometimes extremely so, and the reason for this behaviour is still quite unknown. The sodium or potassium salts of purified conjugates, of whatever chemical type, do not readily take up atmospheric water and I have not found enough organically bound phosphate in some bile salt preparations to account, as lecithin, for their extremely hygroscopic nature.

Further purification of the bile salts depends on their separation on columns or thin layers of silica gel, celite, Sephadex LH-20 or other materials. At present, Sephadex LH-20 columns and thin-layer plates of silicagel are known to have the greatest discriminatory power for separating conjugates. Cronholm et al. (1972) developed a method for separation of various conjugates in chloroform-methanol on Sephadex LH-20 and Almé et al. (1977) have improved this analysis by using ion exchange chromatography on a substituted Sephadex LH-20 column. Some bile, though obtained free from liver or other tissue, does contain polar impurities difficult to separate from bile salts; a sample of such material from fish bile was shown by F. H. Brain and R. W. Baker to be a complex mixture largely composed of hydroxylated fatty acids. Extraction of Sephadex LH-20 itself with chloroform-methanol yields small amounts of substances that may interfere with the identification of conjugates by subsequent infra-red or other spectroscopy. Separation on thin-layer plates of silicagel is effective for some conjugates (Anderson et al., 1974). Commercial silicagel, especially on pre-coated plates, contains considerable amounts of impurities that are also extracted from the gel when separated conjugates are eluted from the plates with methanol. These impurities interfere with subsequent spectroscopy. They can be partially removed from pre-coated plates by successive washings by descending (wick) chromatography with chloroform-methanol-aqueous ammonia, ethyl acetate-acetic acid-water and methanol. Before the methanol washing the plates may be exposed to iodine vapour and they are finally briefly dried at 100°C. Such plates retain their discriminatory power for conjugates and yield less impurity when these are eluted with methanol after separation and visualization with iodine vapour. Other methods for concentrating bile salts from bile, intestinal contents, faeces, urine and blood plasma are given by Eneroth and Sjövall (1971) and Danielsson and Sjövall (1975).

Solvolysis of sulphate esters can be done as follows: the sulphate (up to about 8 mg) is dissolved in 0.1 M aqueous hydrochloric acid (1 ml), sodium chloride (200 mg) is now dissolved in the mixture and butanone (4 ml) is added. The whole is stoppered and kept, with occasional mixing, at 37°C for about 90 hours. If it is desired to estimate sulphate, 0.5 M barium chloride (0.2 ml) may be added and, after about an hour, barium sulphate can be collected on a small sintered glass filter, washed with water and methanol, dried at 100°C and weighed. The bile alcohols can be obtained from the filtrate by extraction (twice) with ethyl acetate. Washing the combined extract with water does not remove unconjugated acids originally present in the bile salts, but the solvolysis does not cleave conjugates of these acids, which are consequently lost. Sulphate esters of bile acids are solvolysed by this procedure, but (possibly) internal esters may also be formed and an alkaline hydrolysis (0.1 M sodium hydroxide, 90—100°C, 1 h) must follow if the free acids are desired. Bile alcohols are formed directly from their sulphate esters by this butanone solvolysis. The solvolysed mixture can then be analysed by GLC—mass spectroscopy and almost all substances present identified (Anderson et al., 1974; Danielsson and Sjövall, 1975).

A careful study of mild methods for solvolysing sulphate esters was made by Goren and Kochansky (1973) and other methods for solvolysis are described, for example by Summerfield et al. (1976) and Eyssen et al. (1976). N.B. Javitt and his colleagues (e.g. Javitt et al., 1976) have developed an analytical process for biological samples that uses 2,2-dimethoxypropane as a reagent for protein precipitation (from blood serum), solvolysis of sulphates and methylation. With mineral acid, 6.8 ml of dimethoxypropane with 1.0 ml of water gives an anhydrous mixture that causes both solvolysis and methylation. If less reagent is used, water remains and neither solvolysis nor methylation occur. An alkaline hydrolysis is included in the method, which ends with quantitative GLC and can be used to determine unconjugated, conjugated, unconjugated sulphated and conjugated sulphated bile acids in the original sample. Van Berge-Henegouwen et al. (1977) describe a combined method for hydrolysis of conjugates followed by solvolysis of sulphate esters in acidified ether, leading without isolation to GLC analysis of bile acid methyl ester acetates: good recoveries are claimed.

As mentioned above, a commercially available enzyme may be used to hydrolyse taurine and glycine conjugates of many C_{24} acids but cannot be applied to taurine-conjugated C_{27} acids, a mild method for solvolysis of which is badly needed.

Total bile salts or bile acids can be estimated with a 3α-hydroxysteroid dehydrogenase available from *Pseudomonas testosteroni* (Iwata and Yamasaki, 1964; Murphy et al., 1970; Schwartz et al., 1974), but this enzyme has rather limited and not very well defined specificity. Under suitable conditions, it quantitatively forms NADH from NAD^+ by removing the hydrogen atoms at C-3 from a 3α-hydroxysteroid. Other bacterial enzymes, specific for hydroxyl groups elsewhere in the steroid ring-system, have been described by Aries and Hill (1970b) and two 7α-hydroxysteroid dehydrogenases from strains of *Escherichia coli* have been applied to the estimation of bile acids and their conjugates having 7α-hydroxyl groups (Haslewood et al., 1973; Macdonald et al., 1974). One of these 7α-hydroxysteroid dehydrogenase preparations has been found to have wide specificity for steroids having a 7α-OH group, but not to act in some cases where sulphate ester groups are also present in the molecule (Haslewood and Haslewood, 1976b). Infra-red spectroscopy in potassium bromide remains an informative and simple method for characterizing separated conjugates, bile alcohols and bile acids and the same is true of nuclear magnetic resonance spectroscopy.

Solvent systems for TLC and reagents for visualizing the separated substances tend to be the particular choice of various laboratories using them. I have found that a freshly made mixture of acetic acid, 3-methyl butyl acetate (amyl acetate) and water, 5:5:2, by volume, is very effective, especially for the separation of conjugates. On a 5×20 cm plate coated with silicagel, all bile alcohol sulphates, taurine and glycine conjugates and unconjugated bile acids run well behind the front in this solvent system. Although its discriminatory power is good it, in common with other systems, does not separate taurine conjugates of chenodeoxycholic and deoxycholic acids; this mixture can be analysed, after elution of the combined spot from the plate, by using first the 3α-hydroxysteroid dehydrogenase and then the 7αhydroxysteroid dehydrogenase enzyme preparations mentioned above (e.g. Macdonald et al., 1974). A good system for separating unconjugated bile acids and alcohols is 2,2,4-trimethylpentane-ethyl acetate-acetic acid, 7:12:3, by volume (cf. Eneroth 1963). For visualizing separated steroids on silicagel plates, the much-used ethanolic solution of phosphomolybdic acid can be, with advantage, replaced by a modification of Usui's (1963) reagent, made by dissolving phosphomolybdic acid (5 g) in acetic acid (100 ml) and carefully adding, with cooling, sulphuric acid (7.5 ml). After stirring at room temperature, this reagent is filtered and kept in the dark. Plates sprayed with it and heated at 100°C show steroids as blue spots on a white or nearly white background, with greater sensitivity than obtained with alcoholic phosphomolybdic acid. Differences in colour shade make it possible to distinguish between some substances (e.g. deoxycholic acid and chenodeoxycholic acid, free or conjugated) on a chromatogram. If the developed plate is first sprayed with a solution of sodium borohydride (5 g/dl in 80% v/v methanol—water) and then, after about 15 min, dried and re-sprayed with Usui's reagent, keto (oxo) derivatives appear on heating, since carbonyl groups have been reduced to hydroxyl in situ (Usui, 1963). In this way it is possible to detect 3α-hydroxy-7-oxo-5β-cholanic acid and its conjugates, as well as other ketones not revealed by phosphomolybdic acid spray reagents alone.

For determination of the concentration of bile salts in blood plasma or serum, GLC or enzymic methods can just about reach to the upper limit of values in healthy persons (e.g. Ali and Javitt, 1970; Murphy et al., 1970; Schwartz et al., 1974). The much more sensitive radio-immunoassay methodology is at present in an early and tentative stage of development but has already proved capable of detecting diurnal and post-prandial variations in blood bile salt concentrations (see Chapter 5).

2.9. Physical Chemistry

Stuart-type molecular models of bile salts show that, whether the configuration at C-5 is α or β, all the $-OH$, $-SO_3^-$ or $-COO^-$ groups (polar groups) can be made to lie in a plane on one side of the molecule with the hydrophobic groups on the other side. With the 5β configuration, the fist-like shape resulting from the L-shaped junction between rings A and B emphasises the hydrocarbon mass on the non-polar side. Fig. 2.1 shows this effect for taurocholate (Formula 2.7).

2.7 Taurocholate.

The 5α-configuration results in a flatter molecule, but also clearly divided into polar and hydrophobic sides, except in the case of myxinol disulphate (Formula 2.8; Fig. 2.2), a primitive bile salt probably not functioning very effectively as a detergent.

2.8 Myxinol disulphate (5α,25ξ-cholestane-3β,7α,16α,26-tetrol 3,26-disulphate).

Thus, in almost all cases, the bile salt molecule is an effective amphiphile and can lie at a fat-water interface with its polar groups in the aqueous and its hydrophobic side in the lipid phase. The distances

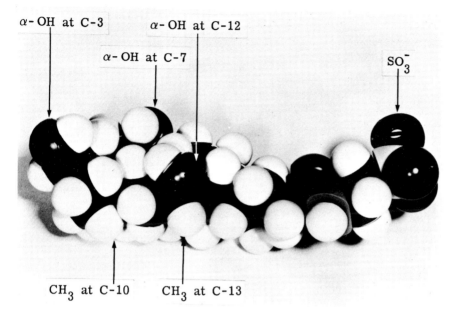

Fig. 2.1. Stuart-type molecular model of taurocholate (Formula 2.7). The polar OH and SO_3^- groups can lie in a plane on one side of the molecule, constituting the hydrophilic part. A water molecule can bridge the OH groups at C-3, C-7 and C-12.

between OH groups at 3α, 7α and 12α are such that a water molecule can easily bridge them: hence hydration in aqueous solution seems very probable.

McBain et al. (1941) were among the first to point out that the bile salts must, like other amphiphilic molecules, exist as micelles in any but very dilute solution. This idea has been extensively examined and given an experimental basis particularly in relation to fat absorption, mainly by A. F. Hofmann, D. M. Small and their co-workers. Comprehensive and well-illustrated reviews are available (e.g. Hofmann, 1968; Small, 1971).

One consequence of micelle formation is, of course, that the nominal dissociation-constants of the bile salts do not truly represent the state of affairs in solution. Removal of a proton creates a negative charge on the micelle and so inhibits subsequent ionization, as in the case of polycarboxylic acids. Thus the actual pH at which a bile salt or bile acid micelle will be ionized to any given extent will be higher than predicted from the Henderson-Hasselbalch equation applied to

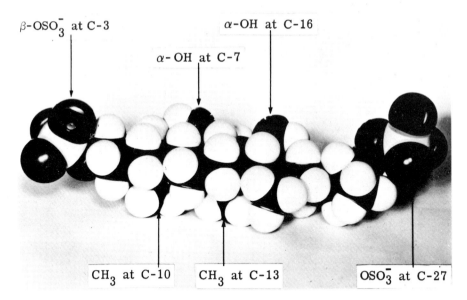

Fig. 2.2. Stuart-type molecular model of myxinol disulphate (Formula 2.8). The SO_3^- and OH groups cannot be spatially separated from the hydrocarbon part of the molecule.

simple solutions and using dissociation constants determined on such solutions. Since the size of micelles varies not only with concentration and pH but also in the presence of ions and of substances capable of forming "mixed micelles", the degree of dissociation actually occurring in vivo is almost unpredictable, except in extreme cases such as that of $-SO_3^-H^+$ groups, whose dissociation-constant is so high that it is hardly conceivable that they could be protonated at any biological pH; i.e. they must always exist as $-SO_3^-$.

Aspects of the physical chemistry of bile salts were well reviewed by Elliott (1971) and Small (1971).

References

Ali, S. S. (1975) Bile composition of four species of amphibians and reptiles. Fed. Proc., 34, 660.
Ali, S. S. and Javitt, N. B. (1970) Quantitative estimation of bile salts in serum. Can. J. Biochem., 48, 1054—1057.
Almé, B., Bremmelgaard, A., Sjövall, J. and Thomassen, P. (1977) Analysis of

metabolic profiles of bile acids in urine using a lipophilic anion exchanger and computerized gas-liquid chromatography-mass spectrometry. J. Lipid Res., 18, 339—362.

Amos, B., Anderson, I. G., Haslewood, G. A. D. and Tökés, L. (1977) Bile salts of the lungfishes *Lepidosiren*, *Neoceratodus* and *Protopterus* and those of the coelacanth *Latimeria chalumnae* Smith. Biochem. J., 161, 201—204.

Anderson, I. G. and Haslewood, G. A. D. (1970) Comparative studies of bile salts. 5α-Chimaerol, a new bile alcohol from the white sucker *Catostomus commersoni* Lacépède. Biochem. J., 116, 581—587.

Anderson, I. G., Haslewood, G. A. D., Oldham, R. S., Amos, B. and Tökés, L. (1974) A more detailed study of bile salt evolution, including techniques for small-scale identification and their application to amphibian biles. Biochem. J., 141, 485—494.

Aries, V. and Hill, M. J. (1970a,b) Degradation of steroids by intestinal bacteria. (a) I. Deconjugation of bile salts. (b) II. Enzymes catalysing the oxidoreduction of the 3α-, 7α- and 12α-hydroxyl groups in cholic acid, and the dehydroxylation of the 7-hydroxyl group. Biochim. Biophys. Acta, 202, 526—543.

Back, P. (1976) Isolation and identification of a chenodeoxycholic acid glucuronide from human plasma in intrahepatic cholestasis. Hoppe—Seyler's Z., 357, 213—217.

Back, P. and Bowen, D. V. (1976) Chemical synthesis and characterization of glucuronic acid coupled mono-, di- and trihydroxy bile acids. Hoppe—Seyler's Z., 357, 219—224.

Cass, O. W., Cowen, A. E., Hofmann, A. F. and Coffin, S. B. (1975) Thin-layer chromatographic separation of sulfated and nonsulfated lithocholic acids and their glycine and taurine conjugates. J. Lipid Res., 16, 159—160.

Cronholm, T., Makino, I. and Sjövall, J. (1972) Steroid metabolism in rats given [1-$^{2}H_{2}$]ethanol. Oxidoreduction of isomeric 3-hydroxycholanoic acid and reduction of 3-oxo-4-cholenoic acid. Eur. J. Biochem., 26, 251—258.

Danielsson, H. and Sjövall, J. (1975) Bile acid metabolism. Ann. Rev. Biochem., 44, 233—253.

Elliott, W. H. (1971) Allo bile acids. In: The Bile Acids. Ed. P. P. Nair and D. Kritchevsky. Vol. 1, pp. 47—93. Plenum Press, New York—London.

Eneroth, P. (1963) Thin-layer chromatography of bile acids. J. Lipid Res., 4, 11—16.

Eneroth, P. and Sjövall, J. (1971) Extraction, purification and chromatographic analysis of bile acids in biological materials. In: The Bile Acids. Eds. P. P. Nair and D. Kritchevsky. Vol. 1, pp. 121—171. Plenum Press, New York—London.

Eyssen, H. J., Parmentier, G. G. and Mertens, J. A. (1976) Sulfated bile acids in germ-free and conventional mice. Eur. J. Biochem., 66, 507—514.

Goren, M. B. and Kochansky, M. E. (1973) The stringent requirement for electrophiles in the facile solvolytic hydrolysis of neutral sulfate ester salts. J. Org. Chem., 38, 3510—3513.

Gordon, B. A., Kuksis, A. and Beveridge, J. M. R. (1963) Separation of bile acid conjugates by ion-exchange chromatography. Can. J. Biochem. Physiol., 41, 77—89.

Hanson, R. F. and Williams, G. (1971) The isolation and identification of 3α,7α-dihydroxy-5β-cholestan-26-oic acid from human bile. Biochem. J., 121, 863—864.

Haslewood, E. S. and Haslewood, G. A. D. (1976a) Preparation of the 3-monosulphates of cholic acid, chenodeoxycholic acid and deoxycholic acid. Biochem. J., 155, 401—404.

Haslewood, E. S. and Haslewood, G. A. D. (1976b) The specificity of a 7α-hydroxy steroid dehydrogenase from *Escherichia coli*. Biochem. J., 157, 207—210.
Haslewood, G. A. D. (1967) Bile Salts. Methuen, London.
Haslewood, G. A. D. (1971) Bile salts of germ-free domestic fowl and pigs. Biochem. J., 123, 15—18.
Haslewood, G. A. D., Ikawa, S., Tőkés, L. and Wong, D. (1978) Bile salts of the Green turtle *Chelonia mydas* (L). Biochem. J., 171, 409—412.
Haslewood, G. A. D., Murphy, G. M. and Richardson, J. M. (1973) A direct enzymic assay for 7α-hydroxy bile acids and their conjugates. Clin. Sci., 44, 95—98.
Haslewood, G. A. D. and Tőkés, L. (1972) A new type of bile salt from *Arapaima gigas* (Cuvier) (family Osteoglossidae). Biochem. J., 126, 1161—1170.
Hayakawa, S. (1973) Microbiological transformations of bile acids. Advances in Lipid Research. Vol. 11, pp. 143—192.
Hofmann, A. F. (1968) Functions of bile in the alimentary canal. In: Handbook of Physiology. Section 6: Alimentary Canal. Ed. C. F. Code. Vol. V, pp. 2507—2533. American Physiological Society, Washington D.C.
Hoshita, T., Hirofuji, S., Sasake, T. and Kazuno, T. (1967) Sterobile acids and bile alcohols. 87. Isolation of a new bile acid, haemulcholic acid from the bile of *Parapristipoma trilineatum* J. Biochem. (Tokyo), 61, 136—141.
Hoshita, T. and Kazuno, T. (1968) Chemistry and metabolism of bile alcohols and higher bile acids. Advances in Lipid Research. Vol. 6. pp. 207—254.
Hoshita, T., Yasuhara, M., Kihara, K. and Kuramoto, T. (1976) Identification of (23S)-5β-cholestane-3α,7α,12α,23,25-pentol in cerebrotendinous xanthomatosis. Steroids, 27, 657—664.
Hsia, S. L. (1971) Hyocholic and muricholic acids. In: The Bile Acids. Eds. P. P. Nair and D. Kritchevsky. Vol. 1, pp. 95—120. Plenum Press, New York—London.
Ikawa, S. and Tammar, A. R. (1976) Bile acids of snakes of the subfamily Viperinae and the biosynthesis of C-23 hydroxylated bile acids in liver homogenate fractions from the Adder, *Vipera berus* (Linn.). Biochem. J., 153, 343—350.
Iwata, T. and Yamasaki, K. (1964) Enzymatic determination and thin-layer chromatography of bile acids in blood. J. Biochem. (Tokyo), 56, 424—431.
Javitt, N. B., Lavy, U. and Kok, E. (1976) Bile acid balance studies in cholestasis. In: The Liver, Quantitative Aspects of Structure and Function. Eds. R. Preisig, J. Bircher and G. Paumgartner. pp. 249—254. Editio Cantor, Aulendorf, West Germany.
Kellogg, T. F. (1973) Bile acid metabolism in gnotobiotic animals. In: The Bile Acids. Eds. P. P. Nair and D. Kritchevsky. Vol. 2, pp. 283—304. Plenum Press, New York—London.
Kihara, K., Yasuhara, M., Kuramoto, T. and Hoshita, T. (1977) New bile alcohols, 5α- and 5β-dermophols from amphibians. Tetrahedron Letters, 8, 687—690.
Kuramoto, T., Kikuchi, H., Sanemori, H. and Hoshita, T. (1973) Bile salts of anura. Chem. Pharm. Bull., 21, 952—959.
Kuramoto, T., Itakura, S. and Hoshita, T. (1974) Studies on the conversion of mevalonate into bile acids and bile alcohols in toad and the stereospecific hydroxylation at carbon atom 26 during bile alcohol biogenesis. J. Biochem. (Tokyo), 75, 853—859.
Lindstedt, S. and Sjövall, J. (1957) On the formation of deoxycholic acid from cholic acid in the rabbit. Bile acids and steroids 48. Acta Chem. Scand., 11, 421—426.

Macdonald, I. A., Williams, C. N. and Mahony, D. E. (1974) A 3α- and 7α-hydroxysteroid dehydrogenase assay for conjugated dihydroxy-bile acid mixtures. Anal. Biochem., 57, 127—136.
Matschiner, J. T. (1971) Naturally occurring bile acids and alcohols and their origins. In: The Bile Acids. Eds. P. P. Nair and D. Kritchevsky. Vol. 1, pp. 11—46. Plenum Press, New York—London.
McBain, J. W., Merrill, R. J. Jr and Vinograd, J. R. (1941) The solubilization of water-insoluble dye in dilute solutions of aqueous detergents. J. Am. Chem. Soc., 63, 670—676.
Murphy, G. M., Billing, B. H. and Baron, D. N. (1970) A fluorimetric and enzymatic method for the estimation of serum total bile acids. J. Clin. Pathol., 23, 594—598.
Myher, J. J., Marai, L., Kuksis, A., Yousef, I. M. and Fisher, M. M. (1975) Identification of ornithine and arginine conjugates of cholic acid by mass spectrometry. Can. J. Biochem., 53, 583—590.
Nair, P. P. (1973) Enzymes in bile acid metabolism. In: The Bile Acids. Eds. P. P. Nair and D. Kritchevsky. Vol. 2, pp. 259—271. Plenum Press, New York—London.
Noma, Y., Noma, Y., Kihara, K., Yashuhara, M., Kuramoto, T. and Hoshita, T. (1976) Isolation of new C_{26} alcohols from bullfrog bile. Chem. Pharm. Bull. 24, 2686—2691.
Okada, S., Masui, T., Hoshita, T. and Kazuno, T. (1962) Isolation of two bile acids from bile of eel. J. Biochem. (Tokyo), 51, 310—311.
Okuda, K. Horning, M. G. and Horning, E. C. (1972) Isolation of a new bile acid, 3α,7α,12α-trihydroxy-5α-cholestan-26-oic acid, from lizard bile. J. Biochem. (Tokyo), 71, 885—890.
Palmer, R. H. (1967) The formation of bile acid sulfates: a new pathway of bile acid metabolism in humans. Proc. Nat. Acad. Sci., 58, 1047—1050.
Parmentier, G. and Eyssen, H. (1975) Synthesis of the specific monosulfates of cholic acid. Steroids, 26, 721—729.
Peric-Golia, L. and Jones, R. S. (1962) Ornithocholanic acids — abnormal conjugates of bile acids. Proc. Soc. Exp. Biol. Med., 110, 327—331.
Peric-Golia, L. and Socic, H. (1968) Free bile acids in sheep. Comp. Biochem. Physiol., 26, 741—744.
Schwarz, H. F., Bergmann, K. V. and Paumgartner, G. (1974) A simple method for the estimation of bile acids in serum. Clin. Chim. Acta, 50, 197—206.
Setogouchi, T., Salen, G., Tint, G. S. and Mosbach, E. H. (1974) A biochemical abnormality in cerebrotendinous xanthomatosis. Impairment of bile acid biosynthesis associated with incomplete degradation of the cholesterol side-chain. J. Clin. Invest., 53, 1393—1401.
Shaw, R. and Elliott, W. H. (1976) Bile acids. XLVIII. Separation of conjugated bile acids by high-pressure liquid chromatography. Anal. Biochem., 74, 273—281.
Shefer, S., Dayal, G., Tint, G. S., Salen, G. and Mosbach, E. H. (1975) Identification of pentahydroxy bile alcohols in cerebrotendinous xanthomatosis: characterisation of 5β-cholestane-3α,7α,12α,24ξ,25-pentol and 5β-cholestane-3α,7α,12α,23ξ,25-pentol. J. Lipid Res., 16, 280—286.
Small, D. M. (1971) The physical chemistry of cholanic acids. In: The Bile Acids. Eds. P. P. Nair and D. Kritchevsky. Vol 1, pp. 249—356. Plenum Press, New York—London.
Summerfield, J. A., Gollan, J. L. and Billing, B. H. (1976) Synthesis of bile acid monosulphates by the isolated perfused rat kidney. Biochem. J., 156, 339—345.
Tamari, M., Ogawa, M. and Kametaka, M. (1976) A new bile acid conjugate, ciliatocholic acid, from bovine gall bladder bile. J. Biochem. (Tokyo), 80, 371—377.
Tammar, A. R. (1974a, b, c) In: Chemical Zoology. Eds. M. Florkin and B. T. Scheer, a, Bile salts in fishes. Vol. VIII, pp. 595—612; b, Bile salts of Amphibia. Vol. IX, pp.

67—76; c, Bile salts in Reptilia. Vol. IX, pp. 337—351. Academic Press, San Francisco.

Usui, T. (1963) Thin-layer chromatography of bile acids with special reference to separation of keto acids. J. Biochem. (Tokyo), 54, 283—286.

Van Belle, H. (1965) Cholesterol, Bile Acids and Atherosclerosis. North-Holland, Amsterdam.

van Berge-Henegouwen, G. P., Allan, R. N., Hofmann, A. F. and Yu, P. Y. S. (1977) A facile hydrolysis-solvolysis procedure for conjugated bile acid sulfates. J. Lipid Res. 18, 118—122.

Yousef, I. M. and Fisher, M. M. (1975) Bile acid metabolism in mammals. VIII. Biliary secretion of cholylarginine by the isolated perfused rat liver. Can. J. Pharmacol., 53, 880—887.

CHAPTER 3

Biosynthesis and artifacts of the enterohepatic circulation

3.1. Biosynthesis: (a) C_{24} 5β bile acids

The subject of this Chapter has been thoroughly reviewed by the most experienced experts and the reader may regard their articles (Danielsson, 1973a; Danielsson and Sjövall, 1975) as the principal sources of original references and opinions at the present time.

All the evidence suggests that in every vertebrate species, cholesterol (or perhaps cholestanol (5α-cholestan-3β-ol) in some cases) is the source of the bile alcohols and C_{27} and C_{24} bile acids.

For the biosynthesis of cholic acid, at least in the common laboratory mammals, man, some other primates and probably in most vertebrates, the stages originally proposed mainly by Sune Bergström and the late Ezra Staple and their colleagues, can still be regarded as indicating the principal route. These are set out in Scheme 3.1.

Some details of the pathways in Scheme 3.1 remain unclear. For example, the enzymes converting 7α-hydroxycholesterol (3.2) into compound 3.3 may have two separable components (a 3β-hydroxy-steroid dehydrogenase and a $\Delta_{5\to4}$ isomerase); it is not certain whether the usual substrate for the important 12α-hydroxylation step is 3.3 or 3.5 and the stages in side-chain degradation (3.7 ⟶ 3.11) are still not well understood. Some of these points are discussed in detail, with references, by Danielsson (1973a).

Part of the evidence leading to Scheme 3.1 has been obtained by feeding isotopically labelled precursors to animals having a biliary fistula and identifying the labelled products in the bile; other evidence has come from the use of perfused isolated whole livers or as a result of incubating presumed intermediates with subcellular fractions from

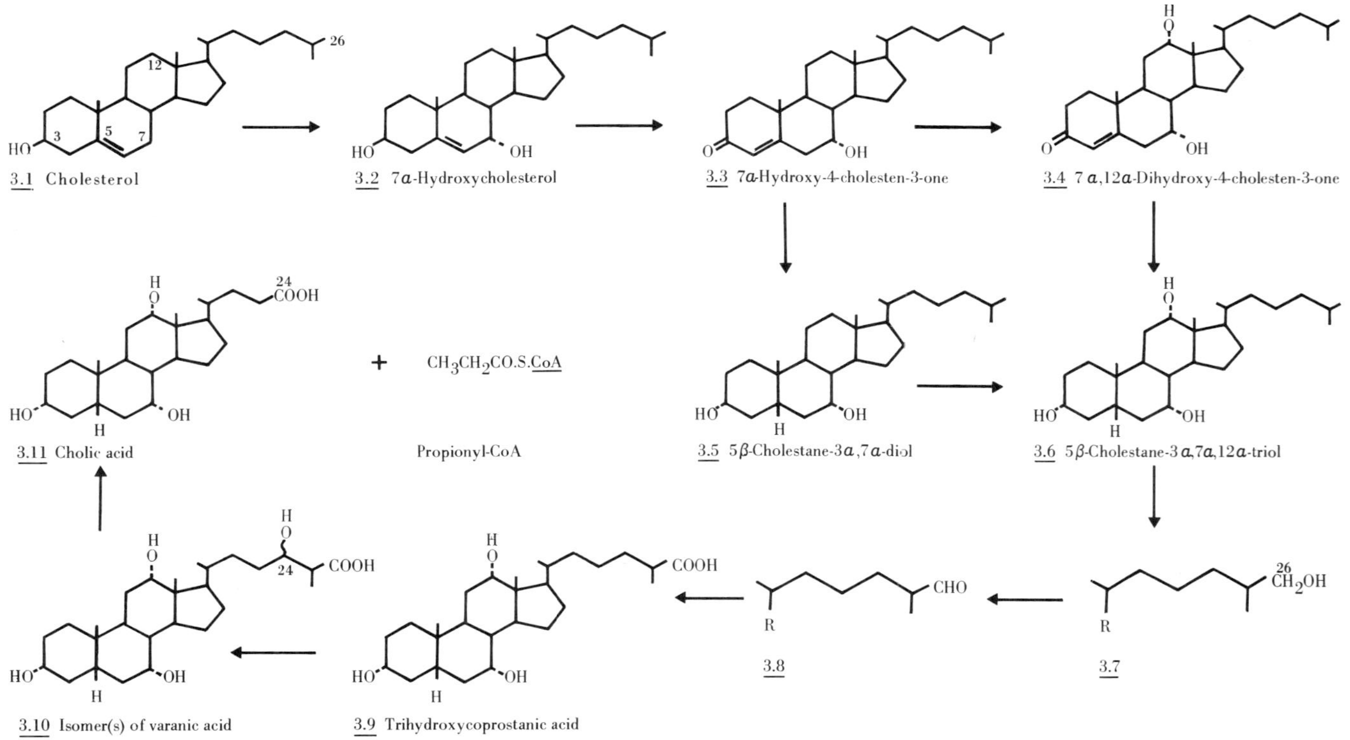

Scheme 3.1. Biosynthesis of cholic acid: probable main route; R = nucleus as 3.6.

liver tissue. The whole sequence cholesterol⟶cholic acid has been carried out with minute yields by liver homogenates. Some of the enzymes concerned seem to be soluble and no doubt it will eventually be possible to characterise them all chemically. The co-factors NAD^+ or $NADP^+$, NADPH and cytochromes of the P_{450} type are involved and enzymic activities have been found in mitochondrial, microsomal and supernatant subcellular fractions.

In two cases at least, namely in the biosynthesis of myxinol in hagfish and of latimerol in the coelacanth *Latimeria* (pages 40 and 84), the 3β-configuration of cholesterol is retained or restored. (There is no evidence to suggest that myxinol or latimerol are secondary bile alcohols.) By studying these animals, particularly the readily available hagfish, it might be possible to throw further light on the earlier steps in Scheme 3.1.

Similarly in some amphibians and reptiles, for example the frog *Discoglossus pictus* and all crocodilians investigated, the side chain although oxidised terminally to COOH can hardly, if at all, be shortened. These reptiles apparently lack the three or four enzymes necessary to complete degradations 3.9 ⟶ 3.11 and their liver tissue fractions might be of interest in comparison with those of more advanced forms carrying out the whole process. Likewise, the alcohol 3.7 and hydroxylated derivatives of it as sulphate esters act as bile salts in certain anurans, which may lack the dehydrogenase responsible for 3.7 ⟶ 3.8 in Scheme 3.1.

These notions are particularly interesting because of recent discoveries that intermediates in or by-products of Scheme 3.1 may accumulate in human disease (see Chapter 5).

The biosynthesis of the second principal C_{24} acid, chenodeoxycholic acid (3.16, Scheme 3.3) is generally believed to be physiologically separated from that of cholic acid at the point at which the 12α hydroxyl group is introduced. Some control, for example directing substance 3.3 of Scheme 3.1 into the endoplasmic reticulum (where it can be hydroxylated at C-12α to give 3.4) or into the cytoplasm, operates here to determine the biosynthesis of one or the other bile acid. If 12α-hydroxylation does not take place, the later stages in Scheme 3.1 could operate in the same way to give chenodeoxycholic acid. Hanson et al. (1976) injected tritium-labelled 7α-hydroxy-4-cholesten-3-one (3.3) intravenously into 8 healthy men. After 1—6 days, a 50% magnesium sulphate solution was put into the duodenum of each subject and a sample of duodenal fluid rich in bile was collected. Comparison of the relative proportions of radioactivity in the derived cholic and chenodeoxycholic acids suggested that some cholic acid might have arisen by a pathway not involving 3.3 (e.g. by earlier 12α-hydroxylation). A careful study on 3 patients with total diversion of bile from the enterohepatic circulation suggested that 5β-cholestane-3α,7α-diol (3.5, Scheme 3.1) may be the "natural substate" for the microsomal 26-hydroxylase, leading to 5β-cholestane-3α,7α,26-triol, a compound that could be a common precursor of cholic and chenodeoxycholic acids. 12α-Hydroxylation would give 3.7 (Scheme 3.1) and so lead to cholic acid, whereas further side-chain oxidation without 12α-hydroxylation would provide chenodeoxycholic acid: the 3α,7α,26-triol was an efficient precursor of both acids in the patients studied (Schwartz et al., 1976). Since the proportion of the two primary bile acids is variable in human bile in health and disease (Chapter 5), it is of practical importance to elucidate the details of the systems governing their formation.

Cholesterol 7α-hydroxylase has been intensively studied. The system requires oxygen, NADPH, cytochrome P_{450}, an NADPH-cytochrome P_{450} reductase and phospholipid. The microsomal activity can be solubilised to some extent. The reaction mechanism is apparently direct replacement of the 7α-H by an OH group (Danielsson and Sjövall, 1975).

The important 12α-hydroxylation step is also microsomal and is stimulated by starvation, which affects the cytochrome P_{450} fraction of a re-constituted system from rat liver. In whole animals, thyroid hormones, bile acids and phenobarbital inhibit 12α-hydroxylation Danielsson and Sjövall, 1975).

Side chain degradation begins with (probably) stereospecific hydroxylation at C-26. Activity is found both in the microsomal and mitochondrial fractions from rat liver and is stimulated by boiled supernatant or NADPH. The microsomal 26-hydroxylase seems to be P_{450}-dependent and shows a preference for 5β-cholestane-3α,7α,12α-triol (3.6) and 5β-cholestane-3α,7α-diol (3.5), the precursors of cholic and chenodeoxycholic acids respectively. The role of the mitochondrial 26-hydroxylase may be important. H. Danielsson and his colleagues are at present making a thorough study of the various hydroxylations in bile acid biosynthesis.

A second mechanism for side-chain oxidation has been postulated by Shefer et al. (1976). These authors noted that Cronholm and Johansson in 1970 had shown that rat liver microsomes could convert

5β-cholestane-3α,7α,12α-triol (3.6) into the 3α,7α,12α,25-tetrol and Björkhem et al. (1975) had demonstrated the same reaction with human material. Shefer et al. (1976) then prepared 5β-cholestane-3α,7α,12α,25-tetrol (3.6a, Scheme 3.2) and showed that it could be converted by liver microsomal fractions, in the presence of oxygen and NADPH, from human patients and also from rats into 5β-cholestane-3α,7α,12α,24R(and S), 25-pentols (3.6b, Scheme 3.2), 5β-cholestane-3α,7α,12α,23ξ,25-pentol and 5β-cholestane-3α,7α,12α,25,26-pentol. In the presence of NAD^+, only the 24R epimer of 3.6b was converted in good yields into cholic acid by the soluble supernatant liver fractions. This work can be interpreted as set out in Scheme 3.2; the C-24 ketone may be involved.

It shows overall that liver cell fractions not containing mitochondria can convert cholesterol into cholic acid, without hydroxylation at C-26. It receives some general support from the discovery in the bile and urine of patients with cerebrotendinous xanthomatosis (CTX) of bile alcohols hydroxylated at C-25 (Table 2.1). Patients with this rare inherited disease have abnormalities in both cholesterol and bile acid biosynthesis. They have in most tissues greatly increased amounts of cholesterol and of cholestanol (5α-cholestan-3β-ol) derived from it. There is a deficiency of bile acids and especially of chenodeoxycholic acid. Salen and Mosbach (1976) describe CTX in detail and attribute the reduced bile acid biosynthesis primarily to defective side-chain oxidation. They and their colleagues have studied five such patients and compared them with others without obvious lipid disease. As a result of more recent experiments with perfused whole laboratory rabbit livers, Cohen et al. (1976) conclude: "Apparently the rabbit forms cholic acid by the classical 26-hydroxylation pathway previously described followed by further degradation of the cholesterol side-chain": it "can adapt and form cholic acid from compounds having a C-25 hydroxyl function but does so to a lesser extent and at a slower rate than the normal pathway". Rabbits make very little chenodeoxycholic acid and Cohen et al. (1976) found that 12α-hydroxylation is very active in this animal, which can hydroxylate 5β-cholestane-3α,7α-diol (3.5) at C-12α after side chain hydroxylation at C-25 or C-26.

There is other evidence for bile acid biosynthesis by routes different from those shown in Scheme 3.1. This Scheme involves the complete formation of a bile acid nucleus before any changes in the side chain, an idea originally proposed by Sune Bergström and his Swedish

54

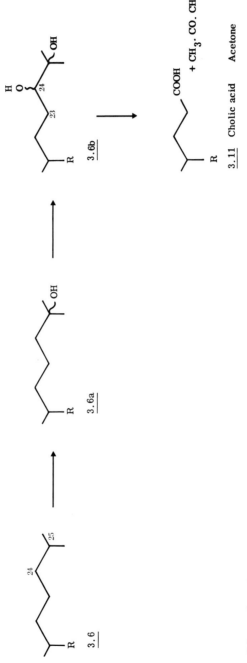

Scheme 3.2. Formation of cholic acid as suggested by Shefer et al. (1976). R = nucleus of 3.6 (Scheme 3.1).

colleagues more than 20 years ago. All the evidence accumulated supports the view that this is normally the chief pathway and it is a striking fact that only a very few undoubtedly primary bile acids and no bile alcohols not having a 7α-OH group have been discovered in nature. A biosynthesis not normally involving early 7α-hydroxylation might perhaps be expected to lead to frequent occurrence of bile salts not so hydroxylated.

However, in 1967 Mitropoulos and Myant reported that rat liver mitochondria together with a supernatant fraction converted ^{14}C-4-cholesterol (3.1) into more polar steroids that included 3β-hydroxy-5-cholesten-26-oic acid (3.13), 3β-hydroxy-5-cholen-24-oic acid (3.14), lithocholic acid (3.15), chenodeoxycholic acid (3.16) and the α- and β-muricholic acids (p. 108) characteristic of rat bile, all apparently as taurine conjugates. To account for these compounds, Mitropoulos and Myant (1967) proposed Scheme 3.3. Although this Scheme leads to chenodeoxycholic acid, 12α-hydroxylation of some (unknown) intermediate would produce cholic acid.

This revolutionary Scheme was at first received with disbelief but in the 10 years since its conception, further evidence has accumulated in its support, normally as a minor route, but perhaps important in certain human diseases and in foetal life. For example, Javitt and Emerman (1970) showed that in rats with a bile fistula both cholesterol and 26-hydroxycholesterol (3.12) were converted into cholic, chenodeoxycholic and muricholic acids and Anderson et al. (1972) found that tritiated 26-hydroxycholesterol (3.12) was quite effective as a precursor of both chenodeoxycholic and cholic acids in six patients with biliary fistulae. From this experiment, it is clear that 12α-hydroxylation can occur after oxidation of the cholesterol side chain. Anderson et al. (1972) did find that 3.12 was not so effective as 7α-hydroxycholesterol (3.2) as a precursor of cholic acid in their patients but preferentially gave chenodeoxycholic acid.

It seems likely that something like Scheme 3.3 operates in foetal bile acid synthesis, for quite large amounts of 3β-hydroxy-5-cholen-24-oic acid (3.14) are found in human meconium (the first intestinal contents of infants) and even in amniotic fluid (Back and Ross, 1973, Délèze et al., 1977).

Scheme 3.3 would certainly account for the hyodeoxycholic acid in germ-free pig bile and for some of the substances found in bile and urine in disease (Chapter 5); it also implies that lithocholic acid could be a primary bile acid.

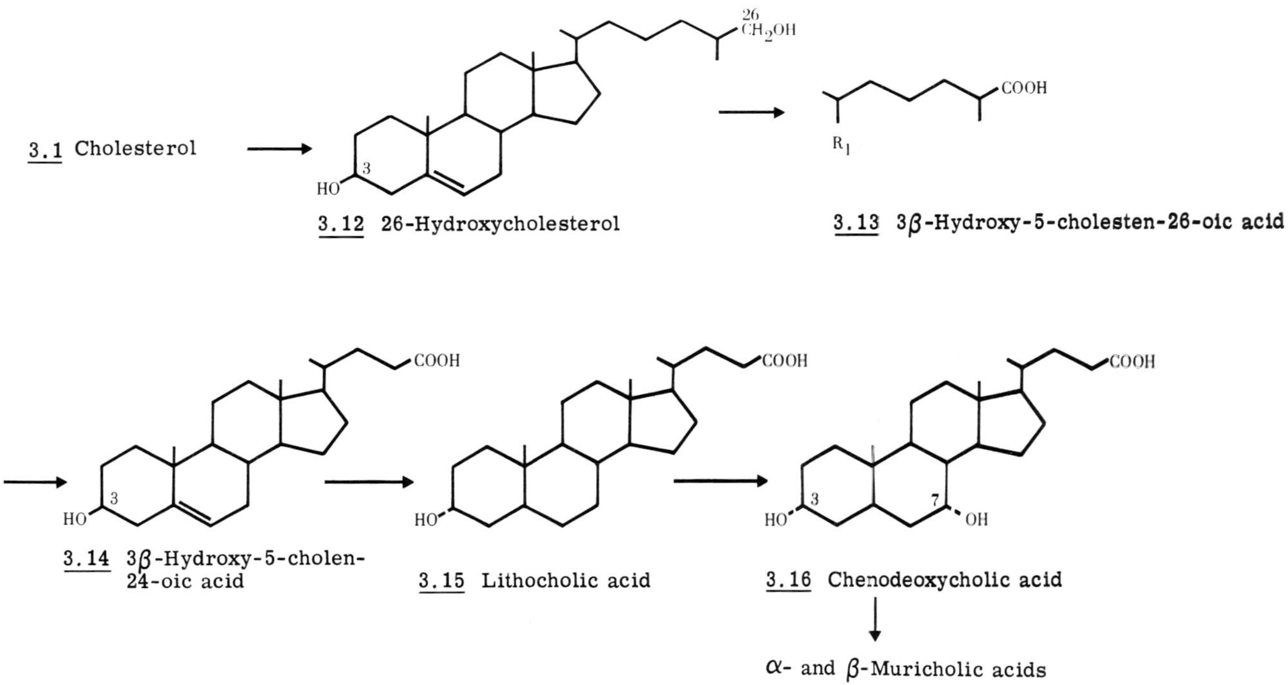

3.1 Cholesterol → 3.12 26-Hydroxycholesterol → 3.13 3β-Hydroxy-5-cholesten-26-oic acid

3.14 3β-Hydroxy-5-cholen-24-oic acid → 3.15 Lithocholic acid → 3.16 Chenodeoxycholic acid

→ α- and β-Muricholic acids

Scheme 3.3. Alternative route for the biosynthesis of bile acids (Mitropoulos and Myant, 1967). R_1 = nucleus as 3.12.

Ayaki and Yamasaki (1970) have shown that 7α-hydroxycholesterol (3.2) could be converted by fortified rat liver mitochondria to 3β,7α-dihydroxy-5-cholen-24-oic acid (3.17, Scheme 3.4) and Yamasaki and Yamasaki (1972) found that the same acid was formed from 7α-hydroxycholesterol in a domestic hen having a bile fistula. Ayaki and Yamasaki (1972) detected 3.17 in the fistula bile of a rat injected with ^{14}C-4-cholesterol and ^{14}C-2-mevalonate and all these experiments were taken as supporting the route to chenodeoxycholic acid shown in Scheme 3.4.

As Danielsson and Sjövall (1975) remark: "the quantitative importance of these different pathways leading to chenodeoxycholic acid has not been established" but there is a good experimental basis for the view that its biogenesis can be separated from that of cholic acid. In some animals at least, chenodeoxycholic acid can be converted directly by 12α-hydroxylation to cholic acid: examples are snakes of the family Boidae (Bergström et al., 1960), a Japanese eel (Masui et al., 1967), cod *Gadus callarias* (Kallner, 1968) and Rainbow trout (Denton et al., 1974).

In reviewing the present position it should be said that it is inherently more likely that experiments with whole animals, even those under restraint and with a biliary fistula, will give results that represent what occurs during normal life. Tissue preparations of any kind carry out reactions that may indeed be possible in vivo, but the disorganisation involved makes it difficult to assess their overall importance. This is true of course of all anabolic pathways, for example those leading from mevalonate to cholesterol or from cholesterol to steroid hormones: nevertheless, so complex and heterogenous are the processes leading to secretion of the membrane-affecting amphiphilic bile salts into the bile canaliculi (Chapter 1) that they may be expected to be easily disrupted in any tissue preparation less complete than healthy whole liver. With bile salt biosynthesis there is the additional complication of the intestinal microflora in the complete animal; for this reason, more experiments might be done on germ-free mammals and birds, whose physiology can be said to be intact (Kellogg, 1973).

The foregoing discussion is about the biosynthesis of what may be called the principal C_{24} primary bile acids in many, if not most, advanced animal forms. There are also a few "unique" primary C_{24} acids whose biogenesis has been studied.

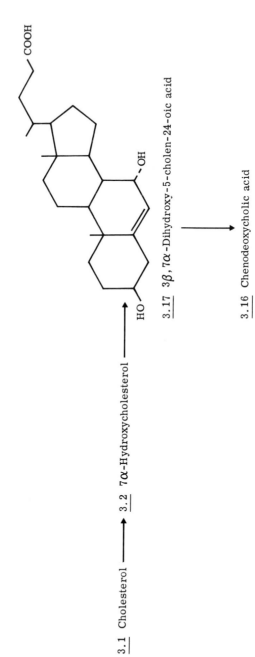

3.1 Cholesterol ⟶ 3.2 7α-Hydroxycholesterol ⟶ 3.17 3β,7α-Dihydroxy-5-cholen-24-oic acid ⟶ 3.16 Chenodeoxycholic acid

Scheme 3.4. Route to chenodeoxycholic acid proposed by Ayaki and Yamasaki (1970).

The pig contains a 6α-hydroxylase system that converts chenodeoxycholic acid into hyocholic acid (p. 114).

Laboratory rats and mice are known to carry out 6β-hydroxylation of chenodeoxycholic acid to make α-muricholic acid: this in turn is converted into its 7β-epimer, β-muricholic acid, which can also be made from ursodeoxycholic acid (4.17, p. 111, Danielsson, 1973a): these acids characterise some rodents at least of subfamily Murinae (p. 108).

It is not known how the C-23 hydroxyl group arises in the phocaecholic acid (4.18) apparently unique to Pinnipedia, but liver preparations from adders, *Vipera berus*, can introduce this group into cholate and deoxycholate and thus make the characteristic C-23 hydroxylated bile acids of certain snakes (p. 100).

Other "unique" C_{24} bile acids, for example haemulcholic acid (with an OH group at C-22) of certain fish, remain to be studied.

The position of some C_{24} acids remains somewhat obscure. Ursodeoxycholic acid (4.17) can possibly be a primary bile acid in rats (Samuelsson, 1959) and 3α-hydroxy-7-oxo-5β-cholanic acid, prominent as its conjugates in the bile of some mammals and birds, may be made without microbial intervention in guinea-pigs.

In summary then: 3α,7α,12α,23ξ-tetrahydroxy-5β-cholanic, cholic, hyocholic, α- and β-muricholic, chenodeoxycholic, hyodeoxycholic, ursodeoxycholic, 3α-hydroxy-7-oxo-5β-cholanic and lithocholic acids can all be primary bile acids although some of them, as described later, are usually secondary.

3.2. Biosynthesis: (b) C_{24} 5α acids

The first of these to be discovered was 5α cholic acid (allocholic acid, Table 2.3). Its presence in the bile of germ-free chicks, rabbits and mice is an absolute proof that it can be primary. Yamasaki and his colleagues were able to demonstrate that a domestic hen having a biliary fistula could make radioactive allocholic acid after intraperitoneal injection of ^{14}C-4-7α-hydroxycholesterol: similar results were obtained with laboratory rabbits and rats (Yamasaki et al., 1972). In a convincing experiment, Hoshita et al. (1968) showed that a microsomal preparation from *Iguana iguana*, which (in common with other lizards of the family Iguanidae) has tauroallocholate as its chief bile salt, would reduce 7α,12α-dihydroxy-4-cholesten-3-one (3.4,

Scheme 3.1) to 5α-cholestane-3α,7α,12α-triol rather than the 5β epimer (3.6, Scheme 3.1) involved in cholic acid biosynthesis. W. H. Elliott (1971) and Danielsson and Sjövall (1975) have reviewed knowledge about 5α-cholanic acids and cite work demonstrating the 7α-hydroxylation of 5α-cholestan-3β-ol (cholestanol) and also its conversion to allocholic acid in rats and the gerbil *Meriones unguiculatus* (Noll et al., 1972). Ali and Elliott (1976) have shown that rats with a bile fistula can make allocholic acid by 12α-hydroxylation of allochenodeoxycholic acid and that the same reaction can be done with rabbit liver microsomes.

A major difference between the biosynthesis of cholic and allocholic acids may have been uncovered by the discovery that rat liver tissue in vivo and in vitro effectively converts 5α-cholestane-3β,7α,26-triol (Formula 3.18) into allocholic acid and allo(5α)chenodeoxycholic acid, whereas the corresponding 5β-compound cannot be 12α-hydroxylated and thus lead to the formation of cholic acid (reference in Danielsson and Sjövall, 1975).

3.18 5α-Cholestane-3β,7α,26-triol (an effective precursor of allocholic acid in rats).

Cholestanol is made by reduction of cholesterol (Salen and Mosbach, 1976) so that animals can make allocholic acid as a primary bile acid in different ways by no means yet fully understood.

A curious fact about allochenodeoxycholic acid is that, unlike allocholic acid, it is uncommon in nature. In germ-free chicks whose chief bile salt is taurine-conjugated chenodeoxycholic acid and which do make allocholic and cholic acids in small amounts, allochenodeoxycholic acid could not be identified with certainty. Ali et al. (1976) found 0.45% by weight of methyl allochenodeoxycholate in the bile acid methyl esters from a lizard *Uromastix hardwickii* (family Agamidae): the methyl esters contained 87% of methyl allocholate. There is no animal known to me in which allochenodeoxycholic acid occurs in more than small amounts. Rats can make it from cholestanol (Elliot, 1971) and so can a gerbil (Noll et al., 1972) and female (but not male) rats have about 8% of their total bile salts as taurine and glycine conjugates of allochenodeoxycholic acid, also partially sulphated (Eriksson et al., 1977). Eyssen et al. (1977) found allocholic and allochenodeoxycholic acid conjugates in both male and female germ-free rats, in which of course these substances must be primary.

3.3. Biosynthesis: (c) C_{27} and C_{28} acids

The best known of these, trihydroxycoprostanic acid (3.9, Scheme 3.1) is a known intermediate in cholic acid biosynthesis in all animals tested, including man (Carey, 1973) and there is no reason to think that it can arise by any other route. The chenodeoxycholic acid analogue, i.e. $3\alpha,7\alpha$-dihydroxy-5β-cholestan-26-oic acid, can be made from cholesterol in man and is efficiently converted to chenodeoxycholic acid, with less than 2% percentage conversion to cholic acid (Hanson, 1971). This experiment again shows that hydroxylation at C-12α can take place even in mammals to some extent when the cholesterol side-chain has been oxidised.

The 5α-epimer of trihydroxycoprostanic acid, $3\alpha,7\alpha,12\alpha$-trihydroxy-5α-cholestane-26(or 27)-oic acid (Table 2.2), is rare in nature but was detected after intraperitoneal injection of tritiated 5α-cholestane-$3\alpha,7\alpha,12\alpha,27$-tetrol into a salamander, *Megalobatrachus japonicus* (Amimoto, 1966).

Nothing is known about the biosynthesis of the $3\alpha,7\alpha,12\alpha,22\xi$-tetrahydroxy-$5\beta$-cholestan-26(or 27)-oic acid of chelonian bile, but isomers of $3\alpha,7\alpha,12\alpha,24\xi$-tetrahydroxy-$5\beta$-cholestan-26(or 27)-oic acid (3.10, Scheme 3.1), are as taurine conjugates, chief bile salts in platynotan lizards (p. 98) and one of them has been isolated together with 3.9, both unconjugated, from a frog *Bombina orientalis* (Kuramoto et al., 1973). This last observation is particularly interesting, for it should be possible to determine the stereochemistry of the *Bombina* acids, in this case substances unaffected in the alkaline hydrolysis necessary to remove taurine. Such a hydrolysis is likely to alter stereochemistry at C-25 and possibly C-24 also. The acids in *Bombina* bile may therefore tell us which isomers of 3.9 and 3.10 are generally used in the biosynthesis of cholic acid.

In an interesting metabolic experiment, Kuramoto et al. (1974) injected ^{14}C-2-mevalonate intraperitoneally into 5 toads, *Bufo vulgaris formosus* (=*Bufo b. formosus*), and after two weeks collected gall-bladder bile. The acid fraction was isolated without alkaline hydrolysis and found to contain radioactive cholic acid and trihydroxycoprostanic acid. The two unsaturated substances, trihydroxy bufosterocholenic acid ($3\alpha,7\alpha,12\alpha$-trihydroxy-5β-cholest-22-ene-24-carboxylic acid; 2.6, p. 33) and $3\alpha,7\alpha,12\alpha$-trihydroxy-5β-cholest-23-en-26(or 27)-oic acid were not labelled. These two unsaturated acids are described by Kuramoto et al. (1974) as "the main bile acids of the toad": their biochemical origin remains a mystery but the latter acid might be metabolically derived from the former (2.6). The cholic acid biosynthesized in this experiment was the first to be found in toad bile and showed that these anurans could complete side chain degradation to give C_{24} bile acids.

3.4. Biosynthesis: (d) bile alcohols

In the work of Kuramoto et al. (1974) mentioned above, the principal toad bile alcohol 5β-bufol (3.19, Scheme 3.5) became labelled. The authors oxidised the isolated bufol with lead tetraacetate to give cholyl methyl ketone (3.20) and formaldehyde (Scheme 3.5). The isolated formaldehyde was not labelled. 3.20 was further oxidised to 25-homocholic acid, with loss of the terminal methyl group and loss of radioactivity corresponding to 0.8 atom of ^{14}C. These results were taken to mean that of the terminal groups of 5β-bufol, —CH_2OH was derived from C-3^1 of mevalonate

$$(^-O\overset{1}{O}C.\overset{2}{C}H_2.\overset{3}{C}(OH).\overset{4}{C}H_2.\overset{5}{C}H_2.OH)$$
$$\underset{3'}{|}\;CH_3$$

and —CH_3 from C-2. This implies that in the biosynthesis of 5β-bufol from cholesterol, oxidation is stereospecific at C-26, as it is in bile acid biosynthesis in all the mammals tested.

In the dogfish experiment described on p. 83, scymnol (the sulphate of which is the chief bile salt in this animal) did not become labelled. This result was attributed to failure of active bile salt bio-

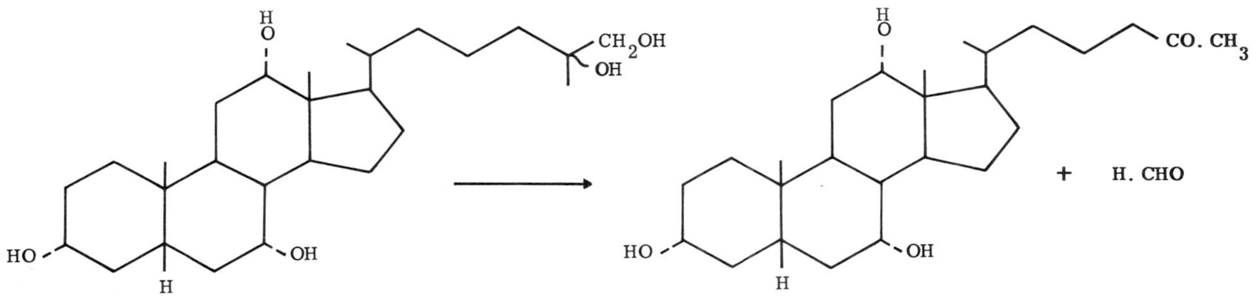

3.19 5 β-Bufol

3.20 Cholyl methyl ketone

Formaldehyde

Scheme 3.5. Conversion of 5β-bufol to cholyl methyl ketone and formaldehyde.

synthesis in captive starving fish; presumably cholic acid biosynthesis here was not under the same physiological control. This study emphasises the necessity of finding out more about the control systems of cold-blooded vertebrates. It cannot be assumed that such animals will constantly carry out biosynthesis as warm-blooded vertebrates normally do, albeit more or less rhythmically.

Takako Masui (1962), in T. Kazuno's laboratory at Hiroshima, injected ^{14}C-4-cholesterol into the peritoneum of three bullfrogs, *Rana catesbeiana*. After two weeks she isolated the bile and hydrolysed it at 160°C for 8 h in 2.5 M sodium hydroxide. These drastic conditions produced radioactive alcohols that were undoubtedly artifacts, but were clearly derived from 5β-ranol (p. 65). Betsuki (1966) similarly injected tritiated 5β-cholestane-3α,7α,12α-triol (3.6, Scheme 3.1) into five specimens of *R. catesbeiana* and, after solvolysis, isolated radioactive 5β-ranol (3.22, Scheme 3.6) and 26-deoxy-5β-ranol (3.21) from the bile. There is no reason to doubt the authenticity of these compounds as natural alcohols (Anderson et al., 1974) and Betsuki suggested that they might arise as shown in Scheme 3.6. If this is correct, C-26 of 3.10 should appear as $^{14}CO_2$ from a carboxyl labelled precursor.

Hoshita (1964) injected ^{14}C-4-cholesterol intraperitoneally into carp, probably *Cyprinus carpio*, and after 12 days isolated radioactive 5α-cyprinol sulphate (4.3, p. 84). This experiment confirms one of mine in which I kept carp for a year on a diet of butchers' meat, supplemented with ox-liver; at the end of this time the bile salts consisted almost entirely of 5α-cyprinol sulphate, as in carp on their usual diet. Since there could have been no more than traces of steroids other than cholesterol in the meat diet, my carp must have been converting endogenous or exogenous cholesterol to their chief bile salt. Neither Hoshita's nor my experiment excluded the enterohepatic circulation. Kouchi (1965) found that goldfish, *Carassius carassius*, could likewise convert ^{14}C-4-cholesterol into radioactive 5α-cyprinol and also cholic acid.

Zoologically and biochemically, the very primitive bile alcohols myxinol, petromyzonol and latimerol are most interesting but nothing is known about their biosynthesis.

In summary, then, we know much about the biosynthesis of the common C_{24} and C_{27} bile acids, but the main metabolic pathways usually used in adult healthy animals are still matters for discussion and further experiment. The idea that two or more routes to a physio-

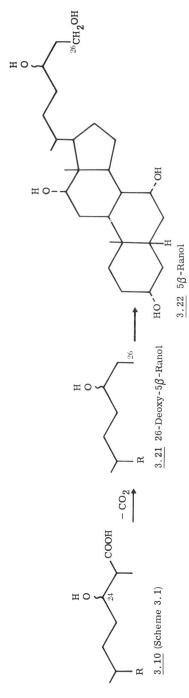

Scheme 3.6. Formation of 5β-ranol and 26-deoxy-5β-ranol, according to Betsuki (1966). R = nucleus as 3.22.

logically functional molecule (in the present case a C_{24} bile acid) are normally used is one not generally supported elsewhere by studies of anabolic vertebrate biochemistry and I find it difficult to accept. However in foetal life, disease or physiological insult the case may be altered. So little is known about the biosynthesis of other bile acids or alcohols that general remarks would be premature.

3.5. Conjugation

Bile alcohols are conjugated with sulphate and I know of no example of their occurrence in free form in bile, except in rare human disease (p. 53). The preliminary work of Bridgwater and Ryan (1957) showed that frog (*Rana temporaria*) liver homogenates incubated with ^{35}S sulphate and 5α-ranol would make a radioactive substance with the mobility on paper chromatograms of natural 5α-ranol sulphate, but these experiments do not seem to have been followed up.

Conjugation of bile acids with sulphate, recently discovered (Chapter 2), occurs by a mechanism not yet investigated, but may, partially at least, be done in the kidney (Summerfield et al., 1976, 1977). Chen et al. (1977) have found and partially characterized an enzyme from rat liver and kidney cytosol fractions that transfers a sulphate groups from 3-phosphoadenosine-5'-phosphosulphate to various bile acids and bile salts; other tissues were inactive.

It has been known for two decades that the conjugation of C_{24} bile acids with taurine or glycine requires the steps set out below.

1. R.COOH + CoA.SH + ATP $\longrightarrow$ R.CO.S.CoA + AMP + pyrophosphate
   Bile acid   CoA   Acyl CoA

2. R.CO.S.CoA + glycine or taurine $\longrightarrow$ R.CO.NH.CH$_2$.COO$^-$ or
   glycine conjugate
   R.CO.NH(CH$_2$)$_2$.SO$_3^-$ + CoA.SH
   taurine conjugate

The enzymes concerned have been found in microsomal and soluble fractions from liver cells; the whole matter has been well reviewed by Nair (1973) in whose article the reader will find an account of the mechanism, nature of the enzymes, methods of assay and application to medicine. Danielsson and Sjövall (1975) give further details and Vessey et al. (1977) showed that, with cholyl CoA, step 2 (above) is catalysed by enzymes in the cytosol fraction from ox liver cells and probably involves separate enzymes for glycine and taurine conjugation. Variations in the proportion of glycine and taurine conjugates occur in response to dietary and hormonal factors and in human disease (Chapter 5). The human infant takes some time (months at least after birth) to attain the adult proportion of conjugates.

3.6. Control of bile salt biosynthesis

In man and the common laboratory mammals at least, the biosynthesis of fresh 5β C_{24} bile acids is thought to be regulated by the concentration of bile salt and bile acid anions in the enterohepatic circulation. This control could be exerted in the liver and be affected by whatever returns to it in the portal circulation from the intestine; i.e. in the intact animal a mixture of bile salts and unconjugated bile acid anions produced by intestinal microorganisms. Of course the liver might re-conjugate such free bile acids before they reach the site of control, which would then be responsive to bile salts alone. If bile is drained away through a fistula or if a bile salt anion-binding resin such as cholestyramine is eaten, there is an increase in the rate of bile salt biosynthesis. The same is true if re-absorption from the intestine is decreased, as for example after removal of the terminal ileum: conversely, feeding bile acids causes a decrease in biosynthesis.

The points of control may be at (1) the 7α-hydroxylation of cholesterol by a 7α-hydroxylase, the first reaction in Scheme 3.1, and (2) the reduction of 3-hydroxy-3-methyl glutaryl CoA (HMG-CoA) to mevalonate, catalysed by HMG-CoA reductase (Scheme 3.7). The enzymic process (2) is believed to govern the rate of cholesterol biosynthesis (e.g. Gregory and Booth, 1975) and its regulation would thus affect the rate of supply of endogenous substrate for process (1).

That the processes of cholesterol and of bile acid biosynthesis might be metabolically linked is suggested, for example, by the studies of Danielsson (1972), who measured six times daily the rate of incorporation of acetate into cholesterol by rat liver slices and the rate of 7α-hydroxylation of cholesterol by the liver microsomal fractions. The two processes showed a closely similar diurnal rhythm, both having minimum rates at noon and maxima between 20.00 h and midnight. Others, quoted by Danielsson (1972), have detected rhythmic variations in both biosyntheses and Mitropoulos and Bala-

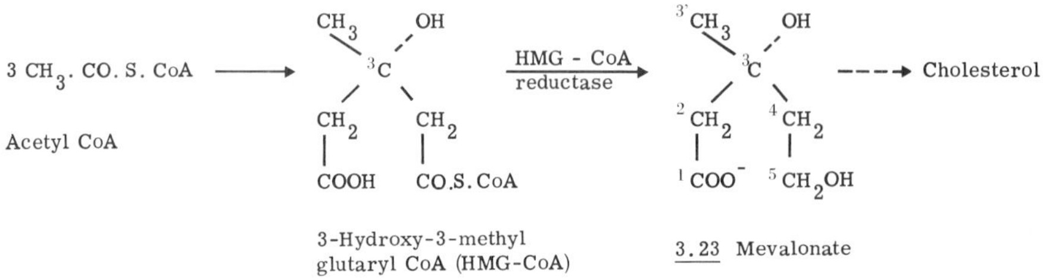

Scheme 3.7. Action of HMG—CoA reductase, a rate-limiting enzyme in cholesterol biosynthesis.

subramaniam (1976) found that these were not abolished in rats by adrenalectomy. Fears and Morgan (1976), however, who regard acetate as "a substrate of doubtful physiological relevance", after studies with ^3H_2O as a substrate to demonstrate cholesterol biosynthesis in rats in various conditions, concluded that "hepatic cholesterogenesis in rats is not subjected to the same degree of diurnal rhythm as has previously been believed". They also agree with earlier expressions of doubt as to the point at which physiological control of cholesterol biosynthesis is most effectively exerted.

Danielsson (1973b) fed to rats a diet containing 1% by weight of the taurine conjugates of cholic, chenodeoxycholic, deoxycholic, hyodeoxycholic or lithocholic acids. The first three, but not the last two, of these conjugates caused, after 3—7 days, a marked inhibition of the liver microsomal cholesterol 7α-hydroxylase activity.

Shefer et al. (1973) fed taurocholate, taurochenodeoxycholate or taurodeoxycholate to whole rats on a stock diet. After 7 days the animals were anaesthetised and bile collected through a cannula for 30—60 min. Then the rats were killed and cholesterol 7α-hydroxylase and HMG-CoA reductase activities measured in the liver microsomal fractions. In all cases the bile salt-fed rats secreted more bile than controls and the bile salt given predominated in the bile. Taurocholate suppressed the activity of both liver microsomal HMG-CoA reductase and cholesterol 7α-hydroxylase but taurochenodeoxycholate affected only the former enzyme, implying "that taurocholate suppresses the synthesis of both cholesterol and bile acids, while taurochenodeoxycholate controls cholesterol biosynthesis but permits a normal rate of bile acid production". The effect of giving taurodeoxycholate was more difficult to interpret, not surprisingly since this substance is efficiently re-oxidised to taurocholate by rat liver (p. 108). Schoenfield et al. (1973) found that after feeding golden Syrian hamsters with cholic or chenodeoxycholic acid (unconjugated) for two weeks, the activities of cholesterol 7α-hydroxylase in the liver microsomal and supernatant fraction and HMG-CoA reductase in the liver microsomal fractions were both significantly less than those from control animals. Chenodeoxycholic acid, however, inhibited cholesterol biosynthesis more than bile acid biosynthesis and the reverse was true of cholic acid, in general agreement with the results of Shefer et al. (1973), mentioned above, with rats. Two studies on the effects of giving bile acids to humans have shown that chenodeoxycholic acid, in contrast to cholic acid, diminishes the output of cholesterol in bile (Adler et al., 1975; LaRusso et al., 1975).

In summary, the evidence is strong that chenodeoxycholic acid, or more probably its conjugates, does have a greater controlling effect on cholesterol biosynthesis than other bile acids tested in rats, hamsters and man. If, as suggested by the work of Mitropoulos et al. (1974), Björkhem and Danielsson (1975), Schwartz et al. (1975) and Normann and Norum (1976) the substrate for bile acid biosynthesis is newly-synthesised cholesterol, we are nearer to an explanation of the apparently close metabolic connection between cholesterol and bile salts.

Giving bile acids or their conjugates to animals affects the secretion of other bile salts; details of experiments of this kind are given by Danielsson and Sjövall (1975), who also review the rest of the literature concerning control of bile salt biosynthesis.

The intestine is very important quantitatively as a source of endogenous cholesterol; indeed it may contribute more by biosynthesis than does the liver. Lutton et al. (1973) studied the turnover of cholesterol in rats fed with bile salts or cholestyramine and concluded that their findings could not be explained if the liver is the principal source of biosynthetic cholesterol. They could reconcile their results with a hypothesis that the intestine is the major source of endogenous cholesterol and that the rate of cholesterol biosynthesis is regulated by the intestinal concentration of bile salts.

3.7. Artifacts of the enterohepatic circulation

These have been mainly investigated (a) by draining off the bile through a fistula and thus excluding the intestine from the circulation, (b) by perfusion of an isolated liver and collection of the bile or (c) by using germ-free animals. The advantage of (a) is that the entire enterohepatic circulation can be restored as required, of (b) that substances can be added which will not be affected by tissues peripheral to the liver and of (c) that an entire physiologically functional animal is employed. Germ-free animals whose bile salts have been studied include laboratory rabbits, rats and mice, domestic fowl and pigs and a baboon. In all cases, oxo (keto) acid and deoxycholic acid conjugates are absent.

None of the methods mentioned above has ever shown that deoxycholic acid can be made by biosynthesis from cholesterol or any of the intermediates in Scheme 3.1. As mentioned on p. 28, conjugated deoxycholic acid disappears from bile drained through a fistula even when, as in rabbits, it normally occurs as a high proportion of the bile salts. Hofmann et al. (1969) found that bile from germ-free rabbits contained only glycine-conjugated cholic (94%), allocholic (1%) and chenodeoxycholic acid (5%). We can therefore take it that deoxycholic acid (2.4) is always made by removal of the 7α-OH group by intestinal bacteria. The mechanism of this reaction in rats having a biliary fistula is a C-6/C-7 dehydration followed by *trans* addition of hydrogen to the double bond formed. Samuelsson (1960) and Aries and Hill (1970) have investigated bacterial enzymes that cause this deoxygenation and found them to be somewhat unstable.

Although hyodeoxycholic acid is usually similarly made from hyocholic acid (Bergström et al., 1959) it also occurs in germ-free pigs (p. 114) and thus can be primary.

Lithocholic acid (3.15, Scheme 3.3) has not been found in germ-free animals but Mitropoulos and Myant (1967) demonstrated that it could be made by bile-fistula rats from 3β-hydroxy-5-cholen-24-oic acid (3.14). Nevertheless, the fairly considerable amounts of lithocholic acid that sometimes occur in bile, faeces or gallstones (in pigs, together with 3β,6α-dihydroxy-5β-cholanic acid) are undoubtedly made by bacterial 7α-deoxygenation of chenodeoxycholic acid (3.16). This situation has important medical implications (Chapter 5).

Ursodeoxycholic acid (4.17) is probably usually secondary and made by reduction at C-7 of 3α-hydroxy-7-oxo-5β-cholanic acid (taurine conjugate, 4.15), which is readily formed from chenodeoxycholic acid by intestinal bacterial 7α-hydroxysteroid dehydrogenase. Rat liver reduces the 7-oxo acid mainly to ursodeoxycholic acid and the guinea-pig makes this also but chiefly chenodeoxycholic acid. As it is now agreed that 3α-hydroxy-7-oxo-5β-cholanic acid can be primary in guinea-pigs, so also can ursodeoxycholic acid (references in Haslewood, 1967).

Escherichia coli and other intestinal bacteria can also readily oxidise (at C-7) cholic acid and its conjugates. The taurine conjugate so formed, namely tauro-3α,12α-dihydroxy-7-oxo-5β-cholanate (or, as A. F. Hofmann might more correctly put it, 3α,12α-dihydroxy-7-oxo-5β-cholan-24-oyl taurate) is converted by germ-free rats to taurocholate and its 7β epimer (Kellogg, 1973). It is thus clear that intestinal microorganisms can cause oxidation or reduction at C-7 and that the liver may modify the product in a way characteristic of the animal species.

As another example, taurodeoxycholate formed from taurocholate in the laboratory rat or mouse intestine is almost completely rehydroxylated by the liver at C-7α, re-forming taurocholate: thus deoxycholic acid conjugates are almost absent from rat and mouse bile. In an analogous way, the livers of many boid snakes (boas and pythons) insert a 16α-OH group into deoxycholate, thus forming the pythocholate characteristic of the bile of these animals (p. 99). Bitocholic acid may similarly arise in viper bile (p. 100). The ability to hydroxylate deoxycholic acid conjugates to re-form trihydroxycholanic acid derivatives might be an adaptation to a change in dietary habits, as suggested on pages 112 and 114.

Probably most of the bile acids found in faeces are not sufficiently absorbed from the lower intestine to reach the liver. Carey (1973) lists 25 human faecal bile acids, most of which are secondary and not normally found in bile.

Bacteria in all parts of the intestine can hydrolyse bile acid conjugates; however, the absorbed bile acids are, in most animals, normally re-conjugated in the liver.

Hayakawa (1973) provides a comprehensive review of work on the changes brought about by microbial action on bile acids.

References

Adler, R. D., Bennon, L. J., Duane, W. C. and Grundy, S. M. (1975) Effects of low dose chenodeoxycholic acid feeding on biliary lipid metabolism. Gastroenterology, 68, 326—334.

Ali, S. S. and Elliott, W. H. (1976) Bile acids. LI. Formation of 12α-hydroxyl derivatives and companions from 5α-sterols by rabbit liver microsomes. J. Lipid Res., 17, 386—392.

Ali, S. S., Farhat, H. and Elliott, W. H. (1976) Bile acids. XLIX. Allocholic acid, the major bile acid of *Uromastix hardwickii*. J. Lipid. Res., 17, 21—24.

Amimoto, K. (1966) Stero-bile acids and bile alcohols. XCII. Metabolism of 27-deoxy-5α-cyprinol in the giant salamander. Hiroshima J. Med. Sci., 15, 225—237.

Anderson, I. G., Haslewood, G. A. D., Oldham, R. S., Amos, B. and Tökés, L. (1974) A more detailed study of bile salt evolution, including techniques for small-scale identification and their application to amphibian biles. Biochem. J., 141, 485—494.

Anderson, K. E., Kok, E. and Javitt, N. B. (1972) Bile acid synthesis in man: metabolism of 7α-hydroxycholesterol — ^{14}C and 26-hydroxycholesterol — ^{3}H. J. Clin. Invest., 51, 112—117.

Aries, V. and Hill, M. J. (1970, a,b) Degradation of steroids by intestinal bacteria. (a) I. Deconjugation of bile salts. (b) II. Enzymes catalysing the oxidoreduction of the 3α, 7α- and 12α-hydroxyl groups in cholic acid, and the dehydroxylation of the 7-hydroxyl group. Biochem. Biophys. Acta, 202, 526—543.

Ayaki, Y. and Yamasaki, K. (1970) In vitro conversion of 7α-hydroxycholesterol to some natural C_{24}-bile acids with special reference to chenodeoxycholic acid biogenesis. (1972) Identification of 3β-7α-dihydroxychol-5-enoic acid in fistula bile of the rat given cholesterol-4-^{14}C and DL-mevalonate-2-^{14}C. J. Biochem. Tokyo, 68, 341—346; 71, 85—89.

Back, P. and Ross, K. (1973) Identification of 3β-hydroxy-5-cholenoic acid in human meconium. Hoppe—Seylers Z., 354, 83—89.

Bergström, S., Danielsson, H. and Garansson, A. (1959) On the bile acid metabolism in the pig. Bile acids and steroids 81. Acta Chem. Scand.,13, 776—783.

Bergström S., Danielsson, H. and Kazuno, T. (1960) Bile acids and steroids 98. The metabolism of bile acids in python and constrictor snakes. J. Biol. Chem., 235, 983—988.

Betsuki, S. (1966) Stero-bile acids and bile alcohols. LXXXIV. The metabolism of trihydroxycoprostane in bull frog. J. Biochem. (Tokyo), 60, 411—416.

Björkhem, I. and Danielsson, H. (1975) 7α-Hydroxylation of exogenous and endogenous cholesterol in rat-liver microsomes. Eur. J. Biochem., 53, 63—70.

Björkhem, I., Gustafsson, J., Johansson, G. and Persson, B. (1975) Biosynthesis of bile acids in man. Hydroxylation of the C_{27}-steroid side chain. J. Clin. Invest., 55, 478—486.

Bridgwater, R. J. and Ryan, D. A (1957) Sulphate conjugation with ranol and other steroid alcohols in liver homogenates from *Rana temporaria*. Biochem. J., 65, 24P.

Carey, Jr., J. B. (1973) Bile salt metabolism in man. In: The Bile Acids. Eds. P. P. Nair and D. Kritchevsky. Vol. 2, pp. 55—82. Plenum Press, New York—London.

Chen, L.-J., Bolt, R. J. and Admirand, W. H. (1977) Enzymatic sulfation of bile salts. Partial purification and characterization of an enzyme from rat liver that catalyzes the sulfation of bile salts. Biochim. Biophys. Acta., 480, 219—227.

Cohen, B. I., Kuramoto, T., Rothschild, M. A. and Mosbach, E. H. (1976) Metabolism of bile alcohols in the perfused rabbit liver. J. Biol. Chem., 251, 2709—2715.

Danielsson, H. (1972) Relationship between diurnal variations in biosynthesis of cholesterol and bile acids. Steroids, 20, 63—72.

Danielsson, H. (1973a) Mechanisms of bile acid biosynthesis. In: The Bile Acids. Eds.

P. P. Nair and D. Kritchevsky, Vol. 2, pp. 1—32. Plenum Press, New York—London.

Danielsson, H. (1973b) Influence of dietary bile acids on formation of bile acids in rat. Steroids, 22, 667—676.

Danielsson, H. and Sjövall, J. (1975) Bile acid metabolism. Ann. Rev. Biochem., 44 233—253.

Délèze, G., Karlaganis, G., Giger, W., Reinhard, M., Sidiropoulos, D. and Paumgartner, G. (1977) Identification of 3β-hydroxy-5-cholenoic acid in human amniotic fluid. In: Bile Acid Metabolism in Health and Disease. Eds. G. Paumgartner and A. Stiehl. pp. 59—62. MTP Press, Lancaster.

Denton, J. E., Yousef, M. K., Yousef, I. M. and Kuksis, A. (1974) Bile acid composition of Rainbow trout, *Salmo gairdneri*. Lipids, 9, 945—951.

Elliott, W. H. (1971) Allo bile acids. In: The Bile Acids. Eds. P. P. Nair and D. Kritchevsky. Vol. 1, pp. 47—93. Plenum Press, New York—London.

Eriksson, H., Taylor, W. and Sjöval, J. (1977) Sex differences in sulfation and excretion of 5α-cholanoic acids in rats. In: Bile Acid Metabolism in Health and Disease. Eds. G. Paumgartner and A. Stiehl. pp. 75—81. MTP Press, Lancaster.

Eyssen, H., Smets, L., Parmentier, G. and Janssen, G. (1977) Sex-linked differences in bile acid metabolism of germfree rats. Life Sci., 21, 707—712.

Fears, R. and Morgan, B. (1976) Studies on the response of cholesterol biogenesis to feeding in rats. Evidence against the existence of diurnal rhythms. Biochem. J., 158, 53—60.

Gregory, K. W. and Booth, R. (1975) Specificity of the effect of dietary cholesterol on rat liver microsomal 3-hydroxy-3-methylglutaryl Coenzyme A reductase activity. Biochem. J., 148, 337—339.

Hanson, R. F. (1971) The formation and metabolism of $3\alpha,7\alpha$-dihydroxy-5β-cholestan-26-oic acid in man. J. Clin. Invest., 50, 2051—2055.

Hanson, R. F., Szczepanik, P. A., Klein, P. D., Johnson, E. A. and Williams, G. C. (1976) Formation of bile acids in man. Metabolism of 7α-hydroxy-4-cholesten-3-one in normal subjects with an intact enterohepatic circulation. Biochim. Biophys. Acta, 431, 335—346.

Haslewood, G. A. D. (1967) Bile Salts. Methuen, London.

Hayakawa, S. (1973) Microbiological transformations of bile acids. Advances in Lipid Research. Vol. 11, pp. 143—192.

Hofmann, A. F., Mosbach, E. H. and Sweeley, C. C. (1969) Bile acid composition of bile from germ-free rabbits. Biochim. Biophys. Acta, 176, 204—207.

Hoshita, T. (1964) Stero-bile acids and bile sterols. LXII. Formation of 5α-cyprinol from cholesterol in carp. Steroids, 3, 523—529.

Hoshita, T., Shefer, S. and Mosbach, E. H. (1968) Conversion of $7\alpha,12\alpha$-dihydroxycholest-4-en-3-one to 5α-cholestane-$3\alpha,7\alpha,12\alpha$-triol by iguana liver microsomes. J. Lipid. Res., 9, 237—243.

Javitt, N. B. and Emerman, S. (1970) Metabolic pathways of bile acid formation in the rat. Mount Sinai J. Med., 37, 477—481.

Kallner, A. (1968) Bile acids in bile of cod, *Gadus callarias*. Hydroxylation of deoxycholic acid and chenodeoxycholic acid in homogenates of cod liver. Bile acids and steroids 200. Acta Chem. Scand., 22, 2361—2370.

Kellogg, T. F. (1973) Bile acid metabolism in gnotobiotic animals. In: The Bile Acids. Eds. P. P. Nair and D. Kritchevsky. Vol. 2, pp. 283—304. Plenum Press. New York—London.

Kouchi, M. (1965) Stero-bile acids and bile alcohols. LXXX. Metabolism of cholesterol in *Carassius carassius*. Hiroshima J. Med. Sci., 14, 195—202.

Kuramoto, T., Kikuchi, H., Sanemori, H. and Hoshita, T. (1973) Bile salts of anura. Chem. Pharm. Bull., 21, 952—959.

Kuramoto, T., Itakura, S. and Hoshita, T. (1974) Studies on the conversion of mevalonate into bile acids and bile alcohols in toad and the stereospecific hydroxylation at carbon atom 26 during bile alcohol biogenetics. J. Biochem. (Tokyo), 75, 853—859.

La Russo, N., Hoffmann, N. E., Hofmann, A. F., Northfield, T. C. and Thistle, J. L. (1975) Effect of primary bile acid ingestion on bile acid metabolism and biliary lipid secretion in gallstone patients. Gastroenterology, 69, 1301—1314.

Lutton, Cl., Mathé, D. and Chevallier, F. (1973) Vitesses des processus de renouvellement du cholestérol contenu dans son espace de transfert, chez la rat. VI Influence de la ligature du cholédoque et de l'ingestion d'acides biliaires ou de cholèstyramine. Biochim. Biophys. Acta, 306, 438—496.

Masui, T. (1962) Stero-bile acids and bile sterols. XXXVII. Formation of bile sterols from cholesterol in bull frog *Rana catesbiana*. J. Biochem. (Tokyo), 51, 112—118.

Masui, T., Ueyama, F., Yashima, H. and Kazuno, T. (1967) Stero-bile acids and bile alcohols. LI. 12α-Hydroxylation of chenodeoxycholic acid in eel liver. J. Biochem. (Tokyo), 62, 650—654.

Mitropoulos, K. A. and Balasubramaniam, S. (1976) The role of glucocorticoids in the regulation of the diurnal rhythm of hepatic β-hydroxy-β-methyl-glutaryl-Coenzyme A reductase and cholesterol 7α-hydroxylase. Biochem. J., 160, 49—55.

Mitropoulos, K. A. and Myant, N. B. (1967) The formation of lithocholic acid, chenodeoxycholic acid and α- and β-muricholic acids from cholesterol incubated with rat-liver mitochondria. Biochem. J. 103, 472—479.

Mitropoulos, K. A., Myant, N. B., Gibbons, G. F., Balasubramaniam, S. and Reeves, E. A. (1974) Cholesterol precursor pools for the synthesis of cholic and chenodeoxycholic acids in rats. J. Biol. Chem., 249, 6052—6056.

Nair, P. P. (1973) Enzymes in bile acid metabolism. In: The Bile Acids. Eds. P. P. Nair and D. Kritchevsky. Vol. 2, pp. 259—271. Plenum Press, New York—London.

Noll, B. W., Walsh, L. B., Doisy, E. A. Jr. and Elliott, W. H. (1972) Bile acids XXXV. Metabolism of 5α-cholestan-3β-ol in the Mongolian gerbil. J. Lipid Res., 13, 71—77.

Normann, Per. T. and Norum, K. R. (1976) Newly synthesized hepatic cholesterol as precursor for cholesterol and bile acids. Scand. J. Gastroenterol., 11, 427—432.

Salen, G. and Mosbach, E. H. (1976) The metabolism of sterols and bile acids in cerebrotendinous xanthomatosis. As Dupont et al. (1976: see References to Chapter 1). pp. 115—153.

Samuelsson, B. (1959) On the metabolism of ursodeoxycholic acid in the rat. Bile acids and steroids. 84. Acta Chem. Scand. 13, 970—975.

Samuelsson, B. (1960) Bile acids and steroids. 96. On the mechanism of the biological formation of deoxycholic acid from cholic acid. J. Biol. Chem., 235, 361—366.

Schoenfield, L. J., Bonorris, G. G. and Ganz, P. (1973) Induced alterations in the rate-limiting enzymes of hepatic cholesterol and bile acid synthesis in the hamster. J. Lab. Clin. Med., 82, 858—868.

Schwartz, C. C., Cohen, B. I., Vlahcevic, Z. R., Gregory, D. H., Halloran, L. G., Kuramoto, T., Mosbach, E. H. and Swell, L. (1976) Quantitative aspects of the conversion of 5β-cholestane intermediates to bile acids in man. J. Biol. Chem. 251 6308—6314.

Schwartz, C. C., Vlahcevic, Z. R., Halloran, L. G., Gregory, D. H., Meek, J. B. and Swell, L. (1975) Evidence for the existence of definitive hepatic cholesterol precursor compartments for bile acids and biliary cholesterol in man. Gastroenterology, 69, 1379—1382.

Shefer, S., Cheng, F. W., Dayal, B., Hauser, S., Tint, G. S., Salen, G. and Mosbach, E. H. (1976) A 25-hydroxylation pathway of cholic acid biosynthesis in man and rat. J. Clin. Invest., 57, 897—903. See also J. Lipid Res., 18, 6—13 (1977).

Shefer, S., Hauser, S., Lapan, V. and Mosbach, E. H. (1973) Regulatory effects of sterols and bile acids on hepatic 3-hydroxy-3-methylglutaryl CoA reductase and cholesterol 7α-hydroxylase in the rat. J. Lipid Res., 14, 573—580.

Summerfield, J. A., Gollan, J. L. and Billing, B. H. (1976) Synthesis of bile acid monosulphates by the isolated perfused rat kidney. Biochem. J., 156, 339—345.

Summerfield, J. A., Cullen, J., Barnes, S. and Billing, B. H. (1977) Evidence for renal control of urinary excretion of bile acids and bile acid sulphates in the cholestatic syndrome. Clin. Sci. Mol. Med., 52, 51—65.

Vessey, D. A., Crissey, M. H. and Zakin, D. (1977) Kinetic studies on the enzymes conjugating bile acids with taurine and glycine in bovine liver. Biochem. J., 163, 181—183.

Yamasaki, H. and Yamasaki, K. (1972) In vivo conversion of 7α-hydroxycholesterol-^{14}C to $3\beta,7\alpha$-dihydroxychol-5-enoic-^{14}C and -4-enoic-^{14}C acids as well as to allocholic -^{14}C acid in the hen. J. Biochem. (Tokyo), 71, 77—83.

Yamasaki, K., Ayaki, Y. and Yamasaki, G. (1972) Allocholic acid, a metabolite of 7α-hydroxycholesterol in the rat and rabbit. J. Biochem. (Tokyo), 71, 927—930.

CHAPTER 4

Distribution of bile salts in the animal kingdom

4.1. Invertebrates*

Invertebrates do not possess a true liver and do not secrete a separate digestive fluid that could be described as bile. They do, however, have surface-tension lowering substances in their alimentary tracts and these substances presumably serve to some extent as do bile salts in vertebrates (e.g. Vonk, 1962; 1969). Two crustaceans have been examined as to the chemical nature of these aids to fat digestion and absorption. The crayfish *Procanbarus clarkii* does not contain known bile acid conjugates (Yamasaki et al., 1965) and the edible crab *Cancer pagurus* has, in its gastric secretion, compounds of fatty acids, sarcosine and taurine that have "detergent" properties (Van den Oord et al., 1965). An example of such a substance is:

$$CH_3(CH_2)_8.CO.N(CH_3).CH_2.CO.NH(CH_2)_2.SO_3^-$$
   decanoyl   sarcosyl   taurine

Arthropods have little or no ability to synthesize sterols (Clayton, 1964; Zandee, 1966; O'Connell, 1970) and perhaps might not be expected to make steroid bile salts. The digestive juices of other invertebrate forms ought to be examined. It is known that some echinoderms can biosynthesize sterols from acetate and mevalonate (see, for example, Smith and Goad, 1975) and this ability is widespread

---

*References to this chapter, when not mentioned, will be found in Haslewood (1967) or in the Appendix.

amongst various groups of invertebrates. P. A. Voogt and his colleagues have examined this question extensively in recent years (Voogt, 1973; 1974) and some work has also come from other sources (e.g. Teshima and Kanazawa, 1974). With the help of Dr. Q. Bone, I examined a collection of digestive caeca from the protochordate lancelet *Branchiostoma (Amphioxus) lanceolatum*. Extracts of these caeca contained substances that ran on thin-layer plates at rates somewhat resembling those of steroid bile salts. However, further study failed to provide evidence that the compounds responsible are steroids.

It seems unlikely that the presence of 5β-cholanic acid in the seeds of a bean, *Abrus precatorius* (Mandava et al., 1974) could have any bearing on digestive function.

4.2. Vertebrates

4.2.1. Cyclostomes

These are the most primitive of the true vertebrates and are the relict highly specialised descendents of jawless fish, the Agnatha. The generally accepted view is that the two extant cyclostome groups, hagfishes and lampreys, are derived from quite different stocks of agnathan fishes and are not closely related.

Morphologically and biochemically, the hagfishes are the more primitive having, for example, a non-myelinated nervous system, innervate hearts, little or no capacity to make antibodies, a primitive type of haemoglobin and an internal inorganic ion composition close to that of the sea-water to which they are confined (Brodal and Fänge, 1963). Representatives of two hagfish genera, *Eptatretus* and *Myxine*, have been shown to have closely similar bile salts, consisting principally of myxinol disulphate (Table 2.1, Formula 2.8, Fig. 2.2). For their size hagfish have large gall-bladders and contain considerable quantities of myxinol disulphate. Molecular models (Fig. 2.2) show that this substance is likely to be a poor amphiphile and perhaps it is for this reason that such large quantities are required by the animals. The digestive problems of hagfish are similar to those of other carnivorous scavengers and predators (Shelton, 1978) and it certainly seems that in their bile salts, too, these animals are making do with molecules less well adapted to their function than is the case in more advanced vertebrates.

Biochemically myxinol disulphate is most primitive, having the complete 5α-cholestane carbon skeleton, with the 3β-OH group of cholesterol itself and no hydroxyl group at C-12. It is also unique; no other disulphate (apart from 16-deoxymyxinol) is known and there is no other (presumably) primary bile salt with a hydroxyl group at C-16.

A second, minor constituent of hagfish bile, 16-deoxymyxinol (Table 2.1) is present (probably also as a disulphate) in such small amounts as to suggest that it may possibly represent myxinol that has escaped 16α-hydroxylation in the course of biosynthesis.

Thus, hagfish bile salts fit well into the picture of a very primitive and specialised vertebrate.

The lamprey *Petromyzon marinus* has a gall-bladder only in the larval (ammocoete) stage. The bile salts seem to comprise just one substance, the sulphate (Formula 4.1) of 5α-petromyzonol (Table 2.1). The same bile salt is probably characteristic of *Petromyzon fluviatalis* and, as Professor T. Kazuno told me, of a Japanese lamprey species also.

4.1 5α-Petromyzonol sulphate (5α-cholane-3α,7α,12α,24-tetrol 24-sulphate).

This compound seems likely to be as good an amphiphile as bile salts generally. Free-living ammocoetes feed on microorganisms and organic detritus and may not have formidable digestive tasks.

Biochemically, petromyzonol is primitive in that it is an alcohol but it does have the 3α,7α,12α-trihydroxy androstane pattern very widely distributed in the animal kingdom. It is also unique in being (with its 5β-epimer, detected in small amounts in lungfish bile) the only known C_{24} alcohol functioning as the basis of a bile salt. Nothing is known about its biosynthesis but it is hardly conceivable that pertromyzonol could arise by reduction of the carboxyl group of the corresponding C_{24} acid (allocholic acid). Lamprey bile salts are thus primitive, unique and not biochemically similar to those of hagfish.

In summary, the nature of their bile salts agrees very well with the systematic position assigned to the two groups of cyclostomes: that they are primitive, specialised and not closely related; also that hagfishes are more primitive than lampreys.

4.2.2. Chondrichthyans

The living cartilagenous fishes can be divided into two groups: the holocephalans (rabbit-fishes, chimaeras) a small group of "fishes which appear to be the specialised survivors of a large and varied group, marine throughout its history" (Patterson, 1965) and the selachians (elasmobranchs) which comprise the sharks and rays. Patterson (1965) reviewed the evidence of relationship between the holocephalans and selachians, remarking that "it has never been doubted that the holocephalans are close to the selachians" but adding "I do not believe that any direct relationship between the two groups is at present demonstrable."

The chemistry of chondrichthyan bile salts, as far as it is known, does clearly point to such a close relationship between the animals and

4.2 5β-Chimaerol (5β-Cholestane-3α,7α,12α,24(+),26-pentol).

also differentiates them from any other vertebrate group. The holocephalan *Chimaera monstrosa* has, as its chief bile salt, the sulphate of 5β-chimaerol (Formula 4.2, Table 2.1) which differs from scymnol (sulphate, Formula 2.2) solely by not having an extra OH group at C-27.

A single biochemical step (27-hydroxylation) is all that is needed to convert 5β-chimaerol into scymnol. Moreover, there is (chromatographic) evidence that scymnol occurs in *Chimaera* bile and 5β-chimaerol has been found in two sharks (Tammar, 1974a). About twelve shark and ray species have scymnol as the chief bile alcohol (Tammar, 1974a).

Scymnol and 5β-chimaerol have not yet been found in any other vertebrate group, so that the evidence from bile salt chemistry emphasises the gap between chondrichthyans and other vertebrates and also the close relationship between the chondrichthyan groups.

Tritiated cholesterol injected into the circulation of captive dogfish (*Scyliorhinus canicula*) led, after some days, to incorporation of tritium into cholic acid (Formula 3.11) but not into scymnol. Cholic acid, earlier found in very small amounts in shark bile, was thus shown to be biosynthesized from cholesterol although in amounts too small for it or its taurine conjugate to be physiologically important as a bile salt. The absence of labelling of the principal bile salt (scymnol sulphate) might be reasonably explained by lack of bile production during starvation entailed by the conditions of captivity (Haslewood, 1969).

4.2.3. Osteichthyans

Osteichthyans: Crossopterygii. The most primitive of the living bony fishes is the coelacanth *Latimeria chalumnae*. Its resemblance to a group of fishes believed until 1938 to have long been extinct led to its description as a "living fossil" and the bile chemistry can be said to be in accordance with this idea. The principal bile salt (see Formula 4.3) in the C-26 (or 27) sulphate of latimerol, which has not been detected in any other animal.

Latimerol is remarkable in having the 3β-OH group of cholesterol, found elsewhere amongst primary bile salts only in hagfish. *Latimeria* bile also contains small and variable amounts of 5α-cyprinol sulphate (Formula 4.3), the 3α epimer of latimerol sulphate, occurring quite widely amongst other bony fishes and amphibians

4.3 5α-Cyprinol sulphate (5α-cholestane-3α,7α,12α,26,27-pentol-26(or 27) sulphate); the 3β epimer is latimerol sulphate.

and characteristic of certain bony fishes regarded as essentially freshwater forms. We can regard latimerol as an evolutionary precursor of 5α-cyprinol, made by an animal in which functional enzymic mechanisms for inverting the 3β-OH group of cholesterol have not been fully elaborated. On this view, *Latimeria* is capable of making this enzymic machinery but only to a very limited extent. Since 5α-cyprinol is characteristic of freshwater forms of fishes, its possession by *Latimeria* (now marine) agrees well with the view that the coelacanths had a long history in fresh water.

Seven samples of *Latimeria* bile have recently been examined by modern techniques. These confirmed the presence of latimerol and 5α-cyprinol and showed in addition minor amounts of 5α-bufol(Formula 4.9), a bile alcohol also found in amphibians and some other primitive osteichthyans (lungfish). The "bile acid" fraction was very small, but might have contained a hydroxylated cholan-24-oic acid. Without more experiments, it cannot be said that such a substance was not derived from the diet of this carnivorous fish; on the other hand its presence might indicate the beginnings of an evolution to C_{24} bile acids (Amos et al., 1977).

Osteichthyans: Dipnoi. Extant lungfishes are assigned to three genera: *Lepidosren* (one species, *L. paradoxa*) from South America, *Neoceratodus* (one species, *N. forsteri*) from Australia and *Protopterus* (four quite closely related species) from Africa. These animals are perhaps the living forms closest to the extinct fish stock from which the amphibians arose. They are also, after *Latimeria*, the most primitive surviving osteichthyans. *Lepidosiren, Neoceratodus* and one species of *Protopterus (P. aethiopicus)* have now been compared in a thorough bile salt examination by present-day techniques (Amos et al., 1977). None of these fishes contains latimerol, a finding which, alone, seems to put them all on a biochemical level above that of *Latimeria*, especially since 5α-cyprinol (which, as remarked above, could be regarded as "modernised" latimerol) is, as its sulphate, a chief bile salt in *Lepidosiren* and *Protopterus*. 27-Deoxy-5α-cyprinol (Formula 4.5) is present as its sulphate in all the lungfishes and is a chief bile salt in *Lepidosiren* and *Neoceratodus*; this bile salt also has the 3α,7α,12α-trihydroxy-5α-androstane ring system, and can be looked upon as the biochemical precursor not only of 5α-cyprinol but also of 5α-chimaerol, whose sulphate (Formula 4.6) replaces 5α-cyprinol as a principal bile salt in *Neoceratodus*. 5α-Bufol (Formula 4.9) sulphate is a chief bile salt in *Lepidosiren* and a minor constituent of *Neoceratodus* bile: it too, can arise by a single hydroxylation (at C-25) step from 4.5. C_{27} alcohol sulphates acting as biles salts in lungfish are listed in Table 4.1.

TABLE 4.1
C_{27} Bile alcohols in lungfish

Lungfish	C_{27} Bile alcohols (as sulphates in bile)	5α-Cholestane hydroxylation pattern
Lepidosiren paradoxa	5α-Bufol[a]	3α,7α,12α,25,26 (or 27)
	5α-Cyprinol[a]	3α,7α,12α,26,27
	27-Deoxy-5α-cyprinol[a]	3α,7α,12α,26 (or 27)
Neoceratodus forsteri	5α-Bufol	
	5α-Chimaerol[a]	3α,7α,12α,24(+),26
	26-Deoxy-5α-chimaerol	3α,7α,12α,24
	27-Deoxy-5α-cyprinol[a]	
Protopterus aethiopicus	5α-Cyprinol[a]	
	27-Deoxy-5α-cyprinol	

[a]Chief bile salt (as sulphate)

Neoceratodus is distinguished from the other lungfish by its ability to hydroxylate at C-24; *Protopterus* apparently cannot insert OH at C-25. *Neoceratodus* has minor amounts of 5α-petromyzonol (sulphate, Formula 4.1) and *Protopterus aethiopicus* also has a little of this substance and of its 5β-epimer. The "bile acid" fractions in the three lungfish are minor constituents of the bile and, allowing for artefacts formed during alkaline hydrolysis, may contain a number of substances with C_{24}, C_{27}, C_{28} and C_{29} steroid skeletons. Some of these might be derived from dietary sterols, but some perhaps represent incipient or abortive bile salt biosynthesis. None is present in quantities suggesting that, alone or as its taurine conjugate, it could make a physiologically important contribution to bile salt digestive function.

In summary, lungfish bile salts are what might be expected of primitive vertebrates not very closely related either to each other or to any other animal types. *Lepidosiren* and *Protopterus* are closer in their bile salt chemistry than either is to *Neoceratodus*. Two of the lungfish (*Neoceratodus* and *Protopterus*) show bile salt affinities with agnathan survivors (lampreys) and the bile of all three genera shows clear resemblances to that of amphibians.

Osteichthyans: Chondrostei. The living fishes of this order (Acipenseriformes) are the sturgeons (Acipenseridae) and paddlefishes (Polyodontidae). The bile salts of three sturgeons and one paddlefish species were found to be very similar (Haslewood and Tammar, 1968). Taurine-conjugated cholic acid was the principal substance but there was also a readily detectable proportion of taurine-conjugated allocholic acid (Table 2.3). These findings suggest an advanced bile salt chemistry with primitive relicts and this notion is supported by the presence in the bile salts of a small proportion (less than 10% by weight) of bile alcohols as monosulphates. The bile alcohol mixture contained both 5α and 5β C_{27} substances: a principal constituent was 5β-cyprinol (Formula 4.4).

5β-Cyprinol sulphate has never been found as a chief bile salt in any fish, although it does play this part in some amphibians. Relict amounts of 5β-cyprinol are, however, quite widespread in bile of groups of bony fishes and this tempts one to suggest that some osteichthyan ancestral stocks (but not those of the crossopterygans, dipnoans or ostariophysans — all of which have 5α-cyprinol) had 5β-cyprinol sulphate as a principal bile salt. This speculation assumes that 5β-cyprinol is biochemically unlikely to be replaced during evolution by 5α-cyprinol and also that the biosynthesis of relict molecules continues to have the physiologically determined selective pressures that, presumably, govern the production of the more abundant, functional, bile salts. Both these assumptions are questionable and it would be unwise to attach too much taxonomic

4.4 5β-Cyprinol (5β-Cholestane-3α,7α,12α,26,27- pentol).

significance to the detailed chemical nature of small amounts of biochemically primitive bile salts occurring in otherwise advanced types of bile. Such considerations do not, nevertheless, invalidate the view that chondrostean bile salts are of an advanced type, with primitive relicts, and this agrees very well with taxonomists' views of the general morphology of these curious fishes.

Osteichthyans: "Holostei". Until recently, the North American bowfin *Amia calva* and garpikes, family Lepisosteidae, were considered to be examples of the holostean level of organisation, differing from teleosts in several anatomical characters. Now, however, *Amia* is regarded as closer to the teleosts than to *Lepisosteus* and fossil holostean fishes (Greenwood, 1975). *Amia calva* bile salts consist, apparently, of taurocholate and Dr. T. Briggs informed me that this is true also of *Lepisosteus osseus*. Thus, the bowfin and garpike do not show the primitive features noticed in chondrostean fish bile.

Osteichthyans: Polypterus and Calamoichthys (Erpetoichthys). These closely related fish genera represent a taxonomic riddle that is far from resolved, little or no help being available from known fossils (Greenwood, 1975). My colleagues and I have examined the bile salts of *Calamoichthys calabricus* and also of *Polypterus congricus*, *P. ornatipinnis* and *P. senegalus*. These animals have principally taurocholate, but also definite proportions of bile alcohol sulphates. Thus, the general picture resembles that shown by the chondrosteans, discussed above, that of advanced fishes with relict primitive characters. I have argued above that it would be unwise at present to attach taxonomic significance to the detailed chemistry of relict bile alcohol mixtures.

Osteichthyans: Teleostei. These, the commonest and most advanced of living fishes are certainly polyphyletic, that is, derived from diverse groups of more primitive stock. P. H. Greenwood and his collaborators have made a number of attempts to classify the teleosts in recent years (e.g. Greenwood et al., 1966; Greenwood, 1975). Their groupings have apparently been made without the help of biochemistry, a handicapped approach since biochemical characters, expressions of DNA sequences, are inevitably indicators of inheritance. This apart, Greenwood is amongst the greatest living authorities on teleostean relationships so that it is encouraging to find that in most respects the bile salt picture does not conflict with his classification.

The superorder Ostariophysi is of outstanding interest in its bile salt chemistry. This enormous group of primarily freshwater fishes is discussed in detail by Greenwood et al. (1966) and divided by Greenwood (1975) as follows:

>   Superorder Ostariophysi
> Series Anotophysi
>   Order Gonorynchiformes
> Series Otophysi
>   Order Cypriniformes
>   Order Siluriformes

Each order is of course further divided into suborders and families and it is amongst the families Cyprinidae and Catostomidae (carps and suckers) of the Cypriniformes that the most primitive teleostean bile salts are to be found. About 13 species of the family Cyprinidae all have 5α-cyprinol sulphate as a principal bile salt and contain only minor amounts of taurine conjugated C_{24} bile acids (Tammar, 1974a). In some species, for example a new *Barbus* and *Varichorhinus* caught in the Upemba uplands of Zaire in 1974 during the Zaire River Expedition, the sulphate of 27-deoxy-5α-cyprinol (Formula 4.5) is also a major bile salt.

C_{27}-Deoxy-5α-cyprinol is a key intermediate in bile alcohol and bile acid biosynthesis, for enzymes are known that can convert it to 5α-cyprinol, 5α-chimaerol and 5α-bufol, as well as to 3α,7α,12α-trihydroxy-5α-cholestanoic acid and allocholic acid itself. This last C_{24}

4.5 27-Deoxy-5α-cyprinol(5α-Cholestane-3α,7α,12α,26-tetrol). (Hydroxylation at C-24 gives 5α-chimaerol; at C-25, 5α-bufol; at C-27, 5α-cyprinol. Oxidation to COOH at C-26 gives 3α,7α,12α-trihydroxy-5α-cholestanoic acid: removal of C-25, C-26 and C-27 and oxidation at C-24 to CH_2OH gives 5α-petromyzonol; oxidation at C-24 to COOH gives allocholic acid.)

acid (the 5α-epimer of cholic acid) is, as its taurine conjugate, one of the minor constituents of cyprinid bile. Cholic acid and chenodeoxycholic acid (3.16) have also been detected, indicating that (in contrast to the primitive osteichthyans containing 5α-cyprinol described earlier) the Cyprinidae have begun bile salt "modernization". It appears therefore that the Cyprinidae have 5α-bile alcohol sulphates as their functioning bile salts, with minor amounts of C_{24} 5α and 5β bile acids probably as taurine conjugates.

The picture amongst the Catostomidae, the suckers of North America, is somewhat different: the bile salts of these fishes seem to be both more varied and also rather more advanced than those of Cyprinidae. In the white sucker, *Catostomus commersoni*, we found what at first seemed to be 5α-cyprinol sulphate, but which further examination proved was 5α-chimaerol sulphate (Formula 4.6: Anderson and Haslewood, 1970).

This substance, replacing in white sucker bile the isomeric and very similar 5α-cyprinol sulphate of carp bile, seems to indicate parallel evolution in cyprinids and catostomids: other suckers, however, suggest that the situation is more complex. Two other *Catostomus* species also have 5α-chimaerol sulphate but two buffalofish, *Ictiobus*, have 5α-cyprinol sulphate and the carpsucker *Carpiodes carpio* taurine-conjugated allocholic acid as principal bile salts (see Appendix). Those suckers which have mainly alcohol sulphate bile

4.6 5α-Chimaerol sulphate (5α,25S-cholestane-3α,7α,12α,24(+), 26-pentol 26-sulphate).

salts also contain small amounts of conjugated C_{24} acids, as do the carps.

The loaches, family Cobitidae, which are placed in the same suborder (Cyprinoidei) as the carps and suckers, probably also have 5α-cyprinol sulphate and C_{24} bile acids; however, since one species only has been examined, this is a tentative conclusion.

Fishes of the order Cypriniformes, but not Cyprinoidei, seem to have biochemically more advanced bile salts. The electric eel *Electrophorus electricus* (family Gymnotidae) has little but taurine-conjugated cholic acid and other C_{24} bile acids and the same is true of a tiger-fish of family Cynodontidae and 3 species of *Distichodus* (family Distichodontidae). Three *Alestes* species (family Charachidae) have high proportions of taurine-conjugated acids, chiefly cheno-deoxycholic acid, with relict traces of bile alcohols.

Examination of the bile of a South American pirhana, probably *Serrasalmus ternetzi*, brought back by an expedition from the Amazon showed only cholate without conjugation: such a remarkable finding requires checking on a fresh specimen collected in conditions certain to exclude post-mortem changes. Fishes assigned to Greenwood's (1975) order Siluriformes have, in general, bile salts biochemically more advanced than those of Cypriniformes, but some species at least do retain primitive molecules. The Japanese Gigi-fish *Pelteobagrus nudiceps* (a small catfish of the family Bagridae) played a historic part in the elucidation of bile salt chemistry. In 1939 Ohta isolated a new bile acid from it, to which he assigned the structure $3\alpha,6\alpha,12\alpha$-trihydroxy-5β-cholan-24-oic acid. Partial synthesis of this acid, twenty years later, proved the incorrectness of this structure and at about the same time Ohta's acid was found in King penguin bile and later shown to be an almost inseparable mixture of cholic and allocholic acids (Anderson and Haslewood, 1962). Thus allocholic acid was recognised in nature and its chemistry elucidated; this discovery then made it possible to find, by infrared spectroscopy, the allocholic acid nucleus in some bile alcohols and so to open up the formerly baffling chemistry of these substances. Another catfish, *Parasilurus asotus* (family Siluridae), has some 5α-cyprinol as well as (mainly) taurine-conjugated cholic and chenodeoxycholic acids.

I found high proportions of unconjugated C_{24} acids in the bile of at least two *Synodontis* species collected on the Zaire River Expedition. I believe that this bile was collected in conditions excluding post-mortem alteration and the observation has now been confirmed with a newly-killed *Synodontis* specimen.

In summary, the Ostariophysi have bile salts varying from the most primitive to the most advanced known in teleosts. Correlation with diet is imperfect: carps and suckers, with most primitive bile salts, are omnivorous but so are some of the catfishes whose bile salts are biochemically advanced. The *Alestes* species referred to above as having mainly taurine-conjugated chenodeoxycholic acid with no more than traces of more primitive substances have the bile salts of vegetarian animals, with some primitive characters.

Finally, the investigator cannot fail to be impressed with the occurrence of the 5α-cyprinol/chimaerol sulphate type of principal bile salt in the following groups of bony fishes: Crossopterygi, Dipnoi, Ostariophysi.

The ostariophysans (Otophysi of Greenwood) all have the Weberian apparatus and must be regarded as monophyletic. The idea that the very obvious bile salt chemical similarities point to an hereditary connection between the three groups of bony fish mentioned is, no doubt, one that would horrify systematists: it should be investigated, nevertheless, particularly to discover how closely the protein molecules acting as enzymes in the biosynthesis from cholesterol of 5α-cyprinol etc. resemble one another in the different fish groups. Comparison of these proteins should make it possible to establish the extent of gene homology in these fishes and hence to throw further light on their relationships.

Another group of teleosts, the family Osteoglossidae, have bile salt molecules of a type not so far found in any other fishes. A principal bile salt of *Arapaima gigas*, an enormous South American freshwater fish of this family, is a sulphate ester of arapaimol-A (Formula 4.7).

4.7 Arapaimol-A (5β-25ξ-cholestane-2β,3α,7α,12α,26-pentol).

The bile also contains the sulphate of the hexol arapaimol-B (with an extra OH group at C-27) and small amounts of the C_{27} acid corresponding to arapaimol-A (COOH instead of CH_2OH at C-26). Taurocholate, too, is a principal bile salt so that the chemistry as a whole is both unique and quite primitive (Haslewood and Tökés, 1972). Greenwood et al. (1968) believe that the Osteoglossiformes could not "possibly have been involved in the ancestry of other teleosteans" and, as far as it goes, the bile salt chemistry is in agreement with this view.

Greenwood's (1975) primitive teleostean cohort Taeniopaedia includes many fishes whose bile salts should be examined, especially since two eels (order Anguilliformes) have some 5β-cyprinol and one has two C_{27} bile acids (Tammar, 1974a). However, two herring species assigned to Greenwood's next most primitive cohort Clupeocephala have, apparently, only C_{24} acids and this is true also of about 45 non-ostariophysan species belonging to his most advanced cohort Euteleostei. An interesting specialization is the occurrence in *Parapristipoma trilineatum* (family Haemulidae, order Perciformes) of the taurine conjugate of haemulcholic acid (3α,7α,22α_F-trihydroxy-5β-cholan-24-oic acid, Table 2.3), a substance found in some other fish (see Appendix). Such "unique" C_{24} acids occur elsewhere in vertebrates and may be indicative of changes in dietary habits during their evolution (see also p. 112).

The deoxycholic acid and keto acids listed by Tammar (1974a) as present in the bile of some teleosts are probably secondarily formed in the intestine during the enterohepatic circulation: these substances are minor biliary constituents, indicating probably that an enterohepatic circulation is not a prominent feature in these fishes: the situation may be different in some examples collected by the Zaire River Expedition.

4.2.4. Amphibians

There are four orders of living amphibians, namely: Apoda (caecilians); Trachystomata (sirens); Urodela (Caudata: newts and salamanders); Anura (Salientia: frogs and toads). It seems likely that the common ancestors of amphibians were fishes of crossopterygian stock and to quote Romer (1966): "frogs and salamanders — and to a less degree the apodans — share a number of seemingly significant common characteristics, strongly suggesting a common ancestry".

Kihara et al. (1977) have found that the sole or principal bile salt in the caecilian *Dermophis mexicanus* is the sulphate of 5β-dermophol (Formula 4.8).

4.8 5β-Dermophol (5β-cholestane-3α,7α,12α,25,26,27-hexol).

Urodeles and anurans have been, and are, objects of great interest especially because of the relative ease by which their development may be followed from the egg through the tadpole or larval stage to the adult form. My colleagues T. Briggs, R. S. Oldham, L. Tökés and I have found definite changes in bile salts as frog tadpole metamorphosis advances. Deuchar (1966) has reviewed the biochemistry of amphibian development and, more recently, D. G. Wallace and his colleagues have compared the serum albumins, fibrinopeptides, haemoglobins and DNA of frogs (Wallace et al., 1971, 1973).

The urodeles examined are the congo "eel" *Amphiuma means*, giant salamander *Megalobatrachus japonicus*, Fire salamander *Salamandra salamandra*, newt *Diemyctylus pyrrohogaster* (Tammar, 1974b; Kihara et al., 1977) and mud-puppy *Necturus* (Ali, 1975). 5α-Dermophol (the 5α-epimer of 4.8) is a minor bile alcohol from *Amphiuma, Diemyctylus* and *Megalobatrachus*; the first two urodeles have the sulphate of 5α-bufol (Formula 4.9) and *Megalobatrachus* that of 5α-cyprinol (Formula 4.2) as chief bile salts, which also occur in lungfish.

4.9 5α-Bufol (5α,25ξ-cholestane-3α,7α,12α,25,26-pentol).

In *Salamandra* there is also present what appears to be a specifically amphibian bile salt, 5α-ranol sulphate (Formula 4.10).

This C_{26} substance and its 5β-epimer occurs in many frogs and toads also, but have not so far been found in any other animal forms. Urodelan bile has C_{27} and C_{24} acids and these are apparently mainly of the 5α-configuration, like the bile alcohols. These preliminary and very limited findings suggest that the bile of urodeles is very similar biochemically to that of anurans, many more of which have been investigated.

4.10 5α-Ranol sulphate (27-nor-5α-cholestane-3α,7α,12α,24ξ-26-pentol 24-sulphate).

It was Japanese work, published in the decade 1930—1940, which first drew attention to the remarkable alcohols and acids in the bile of the bullfrog *Rana catesbeiana* and toad *Bufo b. formosus*. The existence of allo (5α) bile salts had not been suspected and it was not then understood that alkaline hydrolysis was a most unsuitable method for cleavage of sulphate esters. As a result, much of the chemistry proposed by Tayei Shimizu and his colleagues in those early days required revision and, also, some of the substances isolated after alkaline treatment of the bile were artefacts. We owe to those Japanese pioneers our appreciation that amphibian bile salts are in their remarkable diversity probably the most informative of any and the schools at Hiroshima and Yonago continue to make outstanding discoveries, as may be seen from the present text.

A recent survey, employing the most advanced techniques, showed that at least 14 different bile alcohols could be found as their monosulphates in 29 species of anurans: 3α,7α,12α-trihydroxy-5β-cholestan-26-(or 27)-oic acid, taurine-conjugated and as the free ion, was also present in some species together with minor amounts of C_{24} acids, probably mostly as taurine conjugates (Anderson et al., 1974). The general conclusion was that little taxonomic value could be attached to interspecific differences, especially in genera such as Rana and Ptychadena in which every single species examined had quite different bile salts. These differences did not extend to various populations of the same species, however, separate samples of which varied only in the proportions of some of the substances making up their bile salts. In the toad genus *Bufo*, whose chief bile alcohol is the sulphate of 5β-bufol (the 5β-epimer of 4.9), only small differences were found in 5 species, the most remakable being the occurrence in the Japanese *Bufo b. formosus* alone of trihydroxybufosterocholenic acid (Formula 2.5). *Bufo b. formosus* also has a little 5β-dermophol (Formula 4.8; Kihara et al., 1977). Unconjugated bile acids are quite common in anuran bile.

In summary, amphibian bile salts are the most varied of any; they include the sulphates of many bile alcohols, some probably unique to amphibians and others emphasising the relationship between these animals and fishes close to crossopterygian stock. The occurrence of unconjugated bile acid ions as chief bile salts is a remarkable feature of amphibian bile. In a few examples, for example *Salamandra salamandra*, a major bile salt is a taurine-conjugated C_{24} acid but generally the bile salts are biochemically more primitive, consisting mostly of alcohol sulphates and bile acids with the entire cholesterol carbon skeleton. There is so far no indication in bile salts that anurans and urodeles arose from different fish stock.

4.2.5. Reptiles

At this level of organisation, bile alcohols vanish from the picture, at least in healthy animals (they have been found in human bile in certain diseases: see p. 53). Reptilian bile salts are taurine conjugates of C_{27} and C_{24} acids whose distribution accords to some extent with the principal extant reptilian orders as follows: Testudinata (chelonians: tortoises and turtles); Squamata (lizards); Serpentes (snakes); Crocodilia (alligators and crocodiles).

Testudinata (Testudines)

The tortoises and turtles (chelonians) arose from reptilian stock that can be distinguished from the remainder from the oldest appearance of fossil reptiles and this difference is reflected in the bile salts. Until very recently no bile acid had been recognised as being common to chelonians and any other animal group: however, Haslewood et al. (1978) found in the Green turtle *Chelonia mydas* small amounts of cholic and deoxycholic acids. These bile acids can hardly arise from the diet, for the Green turtle feeds almost entirely on seaweed. Tetrahydroxysterocholanic acid (Formula 4.11), conjugated with taurine, is a chief bile salt in at least 6 turtles and tortoises of different families (Tammar, 1974c) and may be considered to be a specifically chelonian bile acid, the ability to make a cholestanoic acid hydroxylated at C-22 having apparently arisen only in this animal group.

The presence of the 7-deoxy derivative of tetrahydroxystereocholanic acid as its taurine conjugate in *Chelonia mydas* is evidence that one C_{27} bile acid at least has an enterohepatic circulation (Haslewood et al., 1978).

4.11 Tetrahydroxysterocholanic acid ($3\alpha,7\alpha,12\alpha,22\xi$-tetrahydroxy-$5\beta$-cholestan-26(or 27)-oic acid).

Squamata (lizards)

One group of lizards, the monitors (Infraorder Platynota, family Varanidae) seem to have as their chief bile salt the taurine conjugate of varanic acid (Formula 4.12) which is, as one of its isomers, an intermediate in the conversion of trihydroxycoprostanic acid to cholic acid (see Chapter 3).

What is probably an isomer of varanic acid was isolated after alkaline hydrolysis of the bile of the beaded poisonous lizard *Heloderma horridum* which, with the closely related *H. suspectum* (Gila monster), constitutes the family Helodermatidae also assigned to the Platynota. This lizard family has features in common with the monitors (Bogert and del Campo, 1956) so that a close relationship between the bile salts is not surprising and it would be interesting to examine other platynotans, for example *Lanthanotus*.

The chemical difficulty about the isomers of varanic acid and other C_{27} acids occurring naturally as taurine conjugates is that because the COOH group is secondary (attached to carbon carrying only one H atom; thus $-CH(CH_3)COOH$), it is somewhat hindered chemically and taurine conjugates require quite drastic conditions for effective alkaline hydrolysis (see also p. 32) which may cause isomerisation, particularly at C-25, so that one is never quite certain that a C_{27} bile acid so obtained occurs as that isomer in the bile. An unconjugated $3\alpha,7\alpha,12\alpha,24\xi$-tetrahydroxy-$5\beta$-cholestanoic acid is a chief bile acid in the frog *Bombina orientalis* (Kuramoto et al., 1973) and the stereochemistry of this should be elucidated completely in case (as seems probable) it is the isomer made in vivo generally and present in the bile of some other animals. It would also be interesting to examine the bile

4.12 Varanic acid ($3\alpha,7\alpha,12\alpha,24\xi$-tetrahydroxy-$5\beta$-cholestan-26(or 27)-oic acid).

of the deadly hybrid *Sampoderma* so well described and illustrated by Bogert and del Campo (1956).

Lizards of all the other families examined so far have taurine conjugates of cholic or allocholic acid as their chief bile salts. The family Iguanidae has, in 8 species, tauroallocholate, with no more than traces of any other bile salt: the same is true of Agamidae (at least 4 species). The glass snake *Ophisaurus* (family Anguidae) has taurocholate and so does at least one lizard (*Lacerta viridis*) of the family Lacertidae.

Serpentes

More than 80 species of snakes, of several families, have been found to have taurine conjugates of 5β, C_{24} acids as their chief bile salts, with no more than minor amounts, in some cases, of 5α, C_{24} acids (Tammar, 1974c). The bile of certain primitive snakes (boas and pythons) and also *Cylindrophis*, all assigned to the Infraorder Henophidia by Underwood (1967), contains an acid, pythocholic acid (Formula 4.13), that throws light on a liver function of these reptiles.

4.13 Pythocholic acid (3α,12α,16α-trihydroxy-5β-cholan-24-oic acid).

Bergström et al. (1960) injected ^{3}H-6α-cholesterol into a python and a boa (species not given) in which a biliary fistula had been established. After saponification the bile yielded labelled cholic and chenodeoxycholic acids. ^{14}C-24-chenodeoxycholic acid similarly yielded labelled cholic acid and ^{14}C-24-deoxycholic acid gave radioactive pythocholic acid. These results mean that, like mammals, boas and pythons can convert cholesterol to cholic and chenodeoxycholic acids, that cholic acid is converted by intestinal microorganisms to deoxycholic acid and, finally, this can be hydroxylated (presumably in the liver) at C-16α to give pythocholic acid.

Thus, the unique character in snakes having pythocholic acid is the ability of the liver to hydroxylate deoxycholate at C-16α: this character has not been detected elsewhere in nature. It is also clear that snakes may have an effective enterohepatic circulation. At least one boa (*Corralus enhydris*) has cholic but not pythocholic acid, therefore lacking either the microorganisms converting cholic acid to deoxycholic acid or the liver enzymes for 16α-hydroxylation of the latter substance.

All the snakes mentioned so far are primitive, having remnants of the limbs of lizards from which they descended. No living lizard has so far been found to have 16α-hydroxylating ability: a lizard with this genetic expression would have to be carefully considered as possibly having a common ancestry with snake precursors.

A second curious character in snakes is hydroxylation at C-23. Both 3α,7α,12α,23ξ-tetrahydroxy-5β-cholanic acid (Formula 4.14) and its 7-deoxy relative, 3α,12α,23ξ-trihydroxy-5β-cholanic acid (bitocholic acid) as their taurine conjugates have been found in snake bile.

4.14 3α,7α,12α,23ξ-tetrahydroxy-5β-cholanic acid (the 7-deoxy compound is bitocholic acid).

The tetrahydroxy acid (Formula 4.14) and bitocholic acid were first discovered in the Gaboon viper *Bitis gabonica* and Puff adder *B. arietans*, together with allocholic and cholic acids in small amounts; 4.14 was later found in a number of vipers (subfamily Viperinae) and also in a few snakes of the family Colubridae.

Ikawa and Tammar (1976) have shown that the liver of the European adder *Vipera berus* can, in vitro, bring about hydroxylation at C-23 of both cholic and deoxycholic acids to form 4.14 and bitocholic acid respectively. Thus, C-23 hydroxylation is a primary (i.e. genetically determined) character and its distribution should have some taxonomic significance.

In his review of snake classification, Underwood (1967) takes account of bile salts as a species character: he is the only zoologist known to me to do so. Whilst there is no systematic difficulty in accommodating C-16α hydroxylation simply as a character of many primitive snakes (but not the "most archaic" *Typhlops*: Underwood, 1967). C-23 hydroxylation is more difficult to understand. Underwood puts both the Viperinae and the Colubridae in the infraorder Caenophidia but this does not necessarily imply any very close relationship. Here again, an examination of the enzymic proteins responsible for C-23 hydroxylation in snakes might yield evidence of gene homology.

Some taurine-conjugated cholic acid is probably present in the bile of all snakes; the other common primary bile acid, chenodeoxycholic acid, was found by Ikawa and Tammar (1976) in 4 vipers. Bergström et al. (1960) showed that boas and pythons, unlike many mammals investigated, could convert chenodeoxycholic to cholic acid (12α-hydroxylation).

The preponderance of 5β, C_{24} bile acids in snakes seems to exclude from their ancestry any lizards having today 5α,C_{24} acids (e.g. Iguania). If it is accepted that 5α is the primitive condition and that 5α→5β may occur in evolution but not the reverse, we should look amongst the ancestors of the 5β lizards (e.g. Scincomorpha, Anguimorpha, Platynota) for the limbed animals more likely to be snake precursors. If the snakes had evolved from 5α lizards we might expect to find more 5α acids in primitive than in advanced snakes, but this does not seem to be the case. Relict amounts of 5α acids occur widely in advanced vertebrates, including man.

The general significance of "unique" 5β, C_{24} bile acids is discussed more fully on pages 112 and 114, but since all snakes are carnivorous it seems unlikely that the occurrence of such bile acids amongst these animals is related to dietary history.

Crocodilia

The Archosauria (crocodiles and alligators) are reptiles whose living forms seems to have diverged but little from their fossil ancestors. To quote Romer (1966): "The crocodiles and alligators and their relatives, the order Crocodilia, have been amongst the least progressive of ruling reptiles. . . . Derived from the theodonts during the course of the Triassic, they have undergone comparatively little modification in later times. In many structural features they are not far from the primitive archosaurian type".

To a remarkable extent crocodilian bile salts accord with this opinion, appearing as they do to represent progression to a certain distance along the path of biosynthesis from cholesterol to cholic acid, and no further. As described in Chapter 3, the main biosynthetic pathway in mammals is cholesterol ⟶ 5β-cholestane-3α,7α,12α-triol ⟶ 3α,7α,12α-trihydroxycoprostanic acid ⟶ 3α,7α,12α,24-tetrahydroxy-5β-cholestan-26-oic acid ⟶ cholic acid. It was shown in 1952 that *Alligator mississipiensis* bile salts are chiefly taurine conjugates of trihydroxycoprostanic acid (Formula 3.9, Table 2.2) and later that this acid is formed from cholesterol and is a precursor of cholic acid in rats and man (Chapter 3). In 1960, T. Briggs injected ^{14}C-4-cholesterol into an alligator having a biliary fistula and isolated radioactive trihydroxycoprostanic acid but no cholic acid from its bile. The bile of 7 other crocodilians seems to resemble that of the alligator, the only other bile acid identified with certainty being 3α,7α-dihydroxycoprostanic acid (Table 2.2) in small amounts. Thus, crocodilian bile is quite primitive and generalised: a halt on the main route of bile salt evolution.

4.2.6. Birds

So little is known about bird bile salts that A. R. Tammar, when requested to review this subject as he has done for fishes, amphibians and reptiles replied that there was insufficient information to justify a new review.

It was for a long time supposed that chenodeoxycholic acid (Formula 3.16) was characteristic of birds, as its name from the Greek χηνός, a goose, suggests. This notion was based on examination of the domestic omnivorous or granivorous fowl and goose and it is not true of birds generally. King penguin (*Aptenodites patagonica*) bile is rather similar to that of some mammals, except that the proportion of allocholic acid is much higher. Tammar (1970) examined 27 bird species by thin-layer chromatography (for conjugates), alkaline hydrolysis and separation of bile acid esters on alumina columns. He found only taurine conjugates and was able to identify some C_{24} acids in almost every case. Cholic and allocholic acids together constituted more than half the bile acids, by weight, in 5 penguin species, 2 eagles and a kite, but chenodeoxycholic acid was the chief bile acid in a vulture and an owl as well as three flamingo, a swan and 2 peafowl species. There does seem to be some correlation with diet: carnivorous birds, like mammals, tending to have more trihydroxy bile acids;

however the correlation is imperfect as the vulture and owl examined by Tammar demonstrate. A remarkable feature of Tammar's work has been the large proportion in some bird species of unidentified bile acids: for example, 378 milligrams of bile acid esters from the hornbill *Bycanistes cylindricus* yielded no substance that could be identified and a crystalline acid from the curassow *Mitu mitu* has so far defied full mass spectral examination. A more determined and sustained chemical attack is required in these and several other cases and might yield results of considerable biological interest. One difficulty in interpretation will be that birds evidently have a good enterohepatic circulation and some bile acids are probably secondary. This problem could in some cases be solved experimentally for germ-free birds (for example, domestic fowl and turkeys) are not difficult to rear to a stage at which bile can be obtained. In germ-free domestic fowl, only chenodeoxycholic, cholic and allocholic acids were found and not the keto acids present in conventional birds (Haslewood, 1971). Not all birds have gall-bladders and the bile of these (e.g. the domestic pigeon *Columba livea*) would have to be obtained by fistula.

Kuramoto and Hoshita (1972) found, in contradiction to earlier work, that the chief bile acids of the kite *Milvus l. lineatus* are trihydroxycoprostanic acid (Formula 3.9) and $3\alpha,7\alpha$-dihydroxy-5β-cholestan-26(or 27)-oic acid. No mammal has more than traces of C_{27} bile acids such as these and this result, together with the common occurrence of quite considerable proportions of allocholic acid in some species, indicates that bird bile is biochemically more primitive than that of mammals. The birds, of course, descended from reptilian stocks quite different from those which gave rise to mammals and this alone is a good reason for wishing to know much more about their bile salts.

4.2.7. Mammals

Simpson (1945) divides the living mammals as follows:

Class Mammalia

Subclass	Protheria
Order	Monotremata
Infraclass	Metatheria
Order	Marsupalia
Infraclass	Eutheria
16 extant orders	

In the following discussion I will generally follow this classification, but a few preliminary remarks about the nature of mammalian bile salts should first be made. At this level of organisation, we find no more than traces of any chemical types other than C_{24} 5β bile acids conjugated with taurine or sometimes, in eutherians only, with glycine. In human disease, however, C_{27} acids have been found in substantial amounts as well as bile alcohols: these are described in their context in Chapter 5. There is a general tendency for carnivorous mammals to have trihydroxy bile acids conjugated with taurine and for vegetarians (except ruminants) to have dihydroxy acids conjugated (in eutherians) with glycine. This association with diet is obvious but incomplete and one is reminded of analogous observations in relation to the mammalian gut pattern made by P. C. Mitchell in 1905 and 1916 and quoted, with further observations, by Le Gros Clark (1962). Mitchell's conclusion that "however great may be adaptive resemblances, the inherited element dominates the structure" may be applied to the bile salts in the sense that there may be a considerable lag between a change in dietary habits and its result, through selection, in altered bile salts. Certainly, short-term feeding has very limited effects as is now well known for human bile. Four dogs fed entirely on a balanced vegetable diet for six weeks showed no differences in the paper chromatographic appearance of their bile salts as compared with four litter-mates fed for the same period entirely on meat (Smyth and Haslewood, 1963). I also fed carp (*Cyprinus carpio*) on butcher's meat for a year without effecting any apparent change, except that increased amounts of red bile pigment (probably bilirubin conjugates) appeared in the bile. However, Sjövall (1959) found that feeding taurine in amounts up to 1.5 g per day increased taurine conjugation in healthy persons, whereas a daily intake of 21 g of glycine had no detectable effect.

The commonest primary bile acid in mammals is cholic acid, as its taurine (or glycine) conjugate (taurocholate: Formula 2.7, Fig. 2.1). In most species, the other primary bile acid is chenodeoxycholic acid (Formula 3.16) also conjugated with taurine (or glycine), but there also occur a few "unique" trihydroxy acids, whose existence might be interpreted as a result of selective pressures brought about by a change in dietary pattern: this idea is more fully discussed on p. 114.

Monotremes

The egg-laying mammals are represented today by the platypus *Ornithorhyncus anatinus* and spiny anteaters (Echidnas), for example *Tachyglossus aculeatus*. Both the species named have been examined and the bile salts found to consist of taurine-conjugated cholic, chenodeoxycholic and deoxycholic acids: no unusual compounds and no glycine conjugates could be detected. These results throw no light on the history of monotremes and give no support to the idea that they have reptilian ancestors different from those of other living mammals (Romer, 1966).

Marsupials

At least 3 kangaroos and an opossum have been found to have cholic, chenodeoxycholic and deoxycholic acids in various proportions, all conjugated with taurine; we have not been able to confirm an earlier report of glycine conjugates in a kangaroo. The strictly vegetarian koala *Phascolarctos cinereus* has no detectable cholic acid and no more than traces of chenodeoxycholic acid. Almost the sole bile salt is 3α-hydroxy-7-oxo-5β-cholanyl taurate (Formula 4.15).

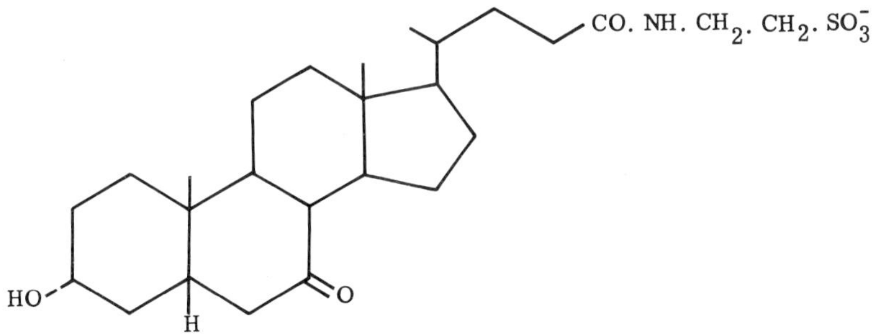

4.15 3α-Hydroxy-7-oxo-5β-cholanyl taurate.

The oxo acid might be formed from chenodeoxycholic acid by intestinal microorganisms having a 7α-hydroxysteroid dehydrogenase (a common enzyme in *Escherichia coli*, for example) or it might be primary, as reported by Peric-Golia and Jones (1961) for guinea-pigs. In any case, we have here a typical vegetarian bile salt type. I have found that the conjugate Formula 4.15 has a mobility on thin-layer chromatograms close to that of taurochenodeoxycholate, which may suggest that its water-fat partition coefficient (one measure of its emulsifying power) is similar also.

Eutherians

Order Insectivora

In the hedgehog *Erinaceus europaeus*, cholic acid was found by methods now superseded: re-examination would no doubt reveal other acids and the nature of conjugation. *Solenodon paradoxus*, by present-day techniques, was found to have taurocholate as the chief bile salt, with traces of conjugated chenodeoxycholic and $3\alpha,12\alpha$-dihydroxy-7-oxo-5β-cholanic acids. Small amounts of glycine conjugates and of unidentified bile acids were also detected (see Appendix).

Order Chiroptera

No bats have been examined.

Order Primates

About 10 monkeys and baboon species and the chimpanzee and orangutan have been compared with man. The three common bile acids, cholic, chenodeoxycholic and deoxycholic acid have been found in almost every case and may be presumed to occur in all. Conjugation is with both glycine and taurine. Superficially, therefore, all these animals are similar.

H. S. Wiggins (1955) made a careful comparison of the bile salts of 4 monkey (*Cercopithecus*) species, the chimpanzee and man. The bile acids were very similar. In the monkey and chimpanzee bile, taurine conjugates constituted 8—54% of bile salts, compared with a range of 10—43% for 13 individual samples of human bile.

O. W. Portman (1962) lists the composition of the bile salts of four monkey species (not fully identified) and the chimpanzee. Except in two *Cebus* monkeys which were pyridoxine-deficient, taurine conjugates predominated but a higher proportion of glycine was found in the Old World *Rhesus* and *Pan* than in the New World *Cebus*, *Lagothrix* and *Saimiri*. The effect of pyridoxine lack in *Cebus* was to increase glycine conjugation but, apart from this, Portman could not detect "any consistent nutritional influence on the distribution of different bile acids" (which were cholic, chenodeoxycholic, deoxycholic and lithocholic acids in this study).

In captive baboons (*Papio anubis* and *P. cynocephalus*) total biliary acids (after hydrolysis) and other bile constituents have been measured, but not individual bile acids or their conjugates (McSherry et al., 1971). These animals "spontaneously" form gallstones and are discussed more fully in Chapter 5. Total bile salt output in Rhesus monkeys after surgery and under restraint has also been measured in various circumstances and this work, too, is considered in Chapter 5, together with the composition of human bile in health and disease.

Order Edentata
Gall-bladder bile of the great anteater *Myrmecophaga tridactyla* was found by Tammar (1970) to contain taurine conjugates only of cholic, chenodeoxycholic and deoxycholic acids; cholic and deoxycholic acids combined made up more than 60% by weight of the isolated bile acids. A trace of allocholic acid was detected. However a sloth, *Choloepus hoffmani*, had glycine as well as taurine conjugates: cholic acid was identified (Tammar, 1970).

Order Pholidota
The scaly anteaters or pangolins, as represented by *Manis tricuspis*, also have only taurine-conjugated bile acids; more than 80% of which consist of cholic acid. *M. pentadactyla* is very similar (Tammar, 1970).

Order Lagomorpha
A great deal of work has been done on the laboratory rabbit, all strains of which are derived from the European *Oryctolagus cuniculus*. This animal is remarkable in that its bile salts normally consist almost entirely of glycine-conjugated deoxycholic acid. Glycine conjugation was shown by Bremer in 1956 to be brought about by the microsomal fraction of rabbit liver cells, which used taurine to a minor extent. As mentioned on p. 28, Linstedt and Sjövall (1957) showed that the rabbit makes glycine-conjugated cholic acid as almost its sole primary bile salt, glycodeoxycholate gradually being formed from this by intestinal microorganisms. In bile acids from two germ-free rabbits, Hofmann et al. (1969) found 93.4 and 94.2% of cholic acid, 2.2 and 0.7% of chenodeoxycholic acid and 4.4 and 5.4% of allocholic acid. Deoxycholic and allodeoxycholic acids were not detected. These results show that in rabbits, as in the domestic fowl (p. 59), allocholic acid may be a primary bile acid. Hofmann et al. (1969) suggested that

in conventional rabbits deoxycholic acid inhibits the biosynthesis of both cholic and chenodeoxycholic acid (which is found also only in minor amounts in this animal). We have here a remarkable example of the dihydroxy bile salts that seem to be characteristic of a vegetarian mammal maintained by the action of its intestinal (or caecal) microorganisms: such a benevolent symbiosis deserves attention. The preponderance of deoxycholic acid may be characteristic also of the European hare *Lepus timidus* and in the California jack-rabbit *L. californicus* both glycine and taurine conjugates are present. The bile acids were found to include 86% of deoxycholic and 12% of cholic acid: chenodeoxycholic and allocholic acids were not detected (Tammar 1970).

Order Rodentia
This huge order contains mammals both omnivorous and vegetarian and includes the laboratory rats, all derived from *Rattus norvegicus*, mice from *Mus muscularis* and the guinea-pig, which now seems to have no wild type. The laboratory rat has no gall-bladder whereas the mouse possesses this organ. Both species are assigned to the subfamily Murinae, so it is perhaps not surprising that they share "unique" bile acids. These are the α- and β-muricholic acids, which are respectively 3α,6β,7α- and 3α,6β,7β-trihydroxy-5β-cholanic acids (Hsia, 1971: Formula 4.16).

4.16 α- and β-muricholic acids.

The major bile acid in both species is normally cholic acid and, in rats at least, chenodeoxycholic and ursodeoxycholic (3α,7β-dihydroxy-5β-cholanic) acids also occur. Rat liver contains enzyme systems that can introduce the 7α-OH group into deoxycholic acid so that this substance, if formed from cholic acid by intestinal microorganisms, is prevented from accumulating in the bile. Germ-free rats excreted about 61% of their faecal bile acids as β-muricholic acid with traces of α-muricholic and about 30% of cholic acid (Kellogg, 1973). Gall-bladder bile acids from germ-free mice included 62% of β-muricholic and 24% of cholic acids with small percentages of allocholic, chenodeoxycholic, α-muricholic and other bile acids, mainly conjugated with taurine, but also to some extent as sulphate esters. In the caecum and large intestine of such mice, allocholic, cholic and chenodeoxycholic acids were sulphated to a very large extent. Conventional mice gall-bladder bile contained ω-muricholic acid (3α,6α,7β-trihydroxy-5β-cholanic acid) which is a secondary bile acid of unknown biogenesis in laboratory rats and mice. A little deoxycholic and lithocholic acids were also present in conventional mouse bile and, in the large intestine and caecum, sulphate esters of bile acids occurred as in germ-free animals (Eyssen et al., 1976, 1977). Glycine conjugation, normally a minor event, can be increased in rats by various feeding regimes and by adrenalectomy (references in Haslewood, 1964).

Simpson (1945) says that the genus *Rattus* "is the most varied genus of mammals, with more than 550 named forms currently recognised". It is a measure of our ignorance that only in *R. norvegicus* amongst these forms is it known that the muricholic acids occur. A survey of the distribution of 6β-hydroxylation followed by an understanding of the proteins concerned should clarify gene homology in *Rattus* and other murine genera. Rats and mice are classified in the rodent suborder Myomorpha, to which are also assigned voles, lemmings and hamsters. The now domesticated Syrian hamster *Cricetus auratus* is of interest because on certain diets it can form gallstones: this work is mentioned in Chapter 5. Hamster bile salts are taurine and glycine conjugates of the three common cholic, chenodeoxycholic and deoxycholic acids.

The rodent suborder Sciuromorpha includes squirrels, ground squirrels and the beaver *Castor canadensis*, in the bile acids of which Tammar (1970) has found 94% by weight of chenodeoxycholic and deoxycholic acids, occurring principally as glycine conjugates but with a little taurine. This is a typical vegetarian type of bile salt mixture.

Guinea-pig bile salts have been the subject of controversy. Most investigators have found them to consist chiefly of glycine-conjugated chenodeoxycholic acid but Peric-Golia and Jones (1961) claimed that adult guinea pigs converted ^{14}C-4-cholesterol into cholic, chenodeoxycholic and 3α-hydroxy-7-oxo-5β-cholanic acid and conjugated these principally with taurine. Danielsson and Einarsson (1964) agree only that the keto acid (taurine conjugate, Formula 4.15) may in this animal be a primary bile acid. Peric-Golia and Danielsson and their colleagues may have been working with different strains of guinea pigs, but I think Danielsson's results to be more credible, especially as they agree so well with the expected picture for vegetarian bile salts.

The bile salts of guinea pigs (rodent suborder Hystricomorpha) may be representative of those of the South American group of rodents, characterised by a high proportion of glycine conjugates. Another example is the coypu *Myocastor coypus* and, through the kindness of Professor M. L. Johnson, my colleagues and I have been able to examine four other members of this group. They are *Capromys pilorides, Geocapromys browni, G. ingrahami* and *Plagiodontia aedium*. All have mainly glycine conjugates and a mixture of bile acids, some of which must be secondarily derived from cholic or chenodeoxycholic acids (see Appendix). One wonders why there is such an obvious preponderance of glycine conjugation in this group of mammals and whether this biochemical feature arose more than once during eutherian evolution.

The West African cutting-grass *Thryonomys swindereanus* resembles the coypu morphologically: indeed G. G. Simpson at one time (1945) placed these rodents together in the superfamily Octodontoidea. Their true relationship is interesting as an example of the biological consequences of the one-time continuity between South America and West Africa, now well explained by the plate tectonic theory of continental drift. Haslewood and Ogan (1959) found that coypu and cutting-grass bile acids were quite similar, but there was no detectable glycine conjugation in the West African rodent. Does this mean that glycine conjugation arose after the separation of the continents?

Order Cetacea

Although whales, porpoises and dolphins do not have gall-badders, it has not proved difficult to obtain whale bile from the stomachs of two whalebone species, the Fin whale, *Balaenoptera physalis* and Sei whale, *B. sibbaldi*. These contain taurine conjugates of cholic and deoxycholic acids; Tammar (1970) could not find glycine conjugates in the former species. A fresh search might reveal chenodeoxycholic acid and the picture we have at present is one of a strictly carnivorous type of bile. There is nothing remarkable here that might give a clue to terrestial cetacean ancestors.

Order Carnivora
About 7 dogs etc. (Canidae), 8 bears (Ursidae), 6 weasels, otters etc., (Mustelidae), the mongoose and kusimanse (Viverridae) and 6 cats (Felidae) have been examined. Except for some of the bears, all have chiefly or solely taurine conjugates of cholic, chenodeoxycholic and deoxycholic acids and some other substances that are probably secondary to these. A long known example of these is ursodeoxycholic acid (Formula 4.17) which was at one time thought to be characteristic of bear bile.

4.17 Ursodeoxycholic acid ($3\alpha,7\beta$-dihydroxy-5β-cholan-24-oic acid).

There was formerly a view that largely vegetarian bears contained mostly glycine conjugates and dihydroxy acids such as ursodeoxycholic and chenodeoxycholic, whereas carnivores like the polar bear had taurine conjugates chiefly of cholic acid. Yamasaki and his colleagues (Kurozumi et al., 1973) have reopened this question and examined by gas-liquid chromatography five different samples of bear bile acids. They did indeed find two types of bile salt, A and B, corresponding to the second and first groups, respectively, mentioned above. The animals are described as Japanese bears, possibly

Selenarctos tibetanus japonicus and Himalyan bears *S. tibetanus* or *Ursus arctos* subspecies. Thus although the new work is zoologically not quite precise, it does confirm that in a single family bile salt variations corresponding to what might be expected from the diet can occur. In all the bear bile examined a fresh ursodeoxycholic acid was a minor constituent, an observation confirmed by I. G. Anderson in my laboratory.

In 2 otters (*Enhydra lutris* and *Lutra sumatrans*) and the Fisher *Martes pennanti*, collected for us by M. L. Johnson, we found taurocholate as the chief bile salt (see Appendix).

Order Pinnipedia (seals, sealions, walruses)
These animals are placed by Simpson (1945) in a suborder, but other authors, for example Sheffer (1958), regard them as an order: "an ancient group — isolated aberrant, unique". Uniqueness is certainly a feature of their bile acids, for in all the species examined (7 seals, 2 sealions and a walrus) phocaecholic acid (Formula 4.18) not discovered elsewhere in nature, has been found.

4.18 Phocaecholic acid ($3\alpha,7\alpha,23\xi$-trihydroxy-5β-cholan-24-oic acid).

The 12α-hydroxy derivative of phocaecholic acid (Formula 4.14, $3\alpha,7\alpha,12\alpha,23\xi$-tetrahydroxy-$5\beta$-cholanic acid) was found in the grey Baltic seal and cholic and chenodeoxycholic acids are probably present generally in pinniped bile salts, all principally as taurine conjugates. It may be speculated that the pinnipeds derive from more herbivorous ancestors whose principal bile acid was chenodeoxycholic acid. When later pinniped precursors took to a more carnivorous diet, it proved biochemically more expedient to develop 23-hydroxylation of chenodeoxycholic acid rather than to increase the proportion of cholic acid, in order to provide the extra trihydroxy bile acids presumably required. The discovery of 23-hydroxylation in any other mammalian group would therefore be of the greatest interest.

Order Tubulidentata
This order is represented by a single species, the Aardvark *Orycteropus afer*, whose bile acids are (percent by weight): allocholic (7), chenodeoxycholic-deoxycholic (combined, 40) and cholic (52): taurine conjugates only could be detected (Tammar, 1970).

Order Proboscidea
Elephants have no gall-bladders; nevertheless it has been reported that *Elephas maximus* contains cholic and deoxycholic acids.

Orders Hyracoidea and Sirenia
No hyraxes or sirenians have been examined, as far as I know. The latter animals (sea-cows, manatees, dugongs) are of special interest because their affinities with terrestrial mammals are obscure. It is possible that their bile salts might be helpful in attemps to elucidate sirenian ancestry.

Order Perrisodactyla
Horses have no gall-bladders; however a little bile expressed from the biliary ducts immediately after the slaughter of a cart-horse was found to contain conjugates of cholic acid. The examination has been repeated on fistula bile with present-day methods (see Appendix).

Order Artodactyla
This large order includes animals without gall-bladders (most of the deer) and also the common domestic farm stock. Amongst this, pigs (infraorder Suina, family Suidae) have been known from the earliest chemically enlightened times to have bile salts different from those of other easily available mammals. The principal bile acid is hyodeoxycholic acid (Formula 4.20, Scheme 4.1), largely but not entirely conjugated with glycine. The corresponding hyocholic acid (Formula 4.19) having an additional OH group at C-6α is a minor constituent; cholic acid is present in traces and the second major bile acid is chenodeoxycholic acid (Formula 3.16). It has been shown that

this can be converted into hyocholic acid by 6α-hydroxylation and it was supposed that hyodeoxycholic acid was a secondary product, formed by bacterial removal of the 7α-OH group (Scheme 4.1).

It was therefore a considerable biochemical surprise when I found quite appreciable amounts of hyodeoxycholic acid in germ-free pigs from two different sources (Haslewood, 1971). Chemical identification was complete: evidently hyodeoxycholic acid can be primary, although most of that present in conventional pigs is no doubt derived as in Scheme 4.1.

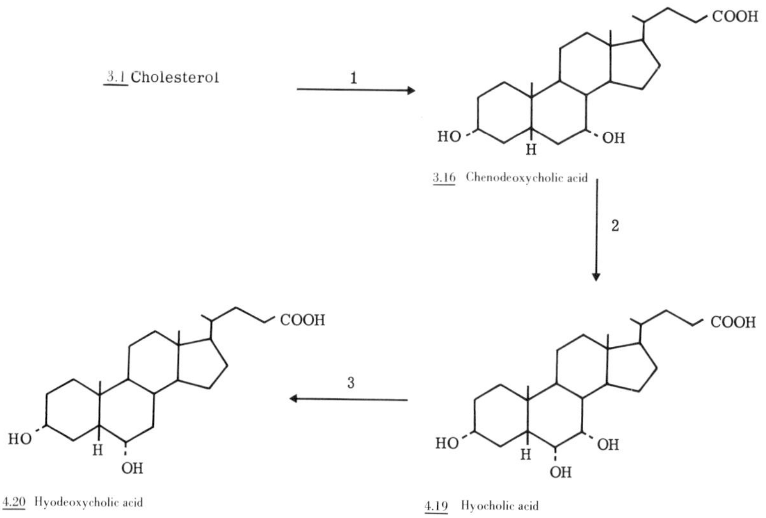

Scheme 4.1. Formation of pig bile acids (Bergström et al., 1959; Samuelsson, 1960). (1) Biosynthesis as in other mammals (Chapter 3); (2) 6α-Hydroxylation in pig liver; (3) Bacterial de-oxygenation in intestine.

It is possible to "explain" the presence of hyocholic acid in the same way as for phocaecholic acid (p. 113), i.e. that when during the course of their evolution the ancestral pigs "required" a trihydroxy bile acid, it was then biochemically more expedient to develop 6α-hydroxylation than to re-activate cholic acid biosynthesis. If that had been the case, there must have been a period when hyocholic acid was a major bile acid and it seems likely that the present-day situation where it is a minor constituent has a more complicated evolutionary origin.

Pigs from the wild (the Abyssinian wild pig) and the semi-feral Nigerian pig (all no more than sub-species or varieties of *Sus scrofa*) have closely similar bile salts (Ogan, 1960). Another member of the subfamily Suinae, the wart hog *Phocochoerus aethiopicus*, also has a little hyodeoxycholic acid but chiefly cholic and chenodeoxycholic acids, all conjugated with both taurine and glycine (Tammar, 1970). We should be justified in saying that, on the basis of findings in two species, 6α-hydroxylation is characteristic of the Suinae; others of this small subfamily await examination. The well-known 3α-hydroxy-6-oxo-cholanic acid of pig bile is secondary; it is not present in germ-free animals. New World pigs, the peccaries (subfamily Tayassuinae) are thus of great interest. They have no gall-bladder and bile probably would have to be obtained from a fistula.

The hippopotamus, infraorder Ancodonta, also has mainly glycine conjugates and the bile acids (percent by weight) included cholic acid (30) and a chenodeoxycholic-deoxycholic acid mixture (47). 23% of the bile acid mixture was not identified but did not, apparently, include 6α-hydroxylated acids (Tammar, 1970).

The remainder of the artiodactyls examined have been shown to have cholic, chenodeoxycholic and deoxycholic acids with small amounts, in most cases, of allocholic acid and traces of keto acids. A curious fact is that in the entirely vegetarian ruminant sheep, goats and cows, the bile salts contain a considerable percentage of taurocholate, in contrast to those of most herbivorous eutherians. A possible partial explanation, for sheep at least, is that they can apparently absorb fats at an intestinal pH lower than that found in non-ruminants; for this purpose taurine conjugates with the low pK would have an obvious advantage (Scott and Lough, 1971).

Assessing the general bile salt picture in mammals, we may conclude that although in healthy animals only a limited number of C_{24} 5β-cholanic acids are, as their conjugates, acting as bile salts, nevertheless the differences found may indeed have some taxonomic value especially in shedding light on ancestral forms.

References

Ali, S. S. (1975) Bile composition of four species of amphibians and reptiles. Fed. Proc., 34, 660.

Amos, B., Anderson, I. G., Haslewood, G. A. D. and Tökés, L. (1977) Bile salts of the lungfishes *Lepidosiren, Neoceratodus* and *Protopterus* and those of the coelacanth *Latimeria chalumnae* Smith. Biochem. J., 161, 201–204.

Anderson, I. G. and Haslewood, G. A. D. (1962) Comparative studies of 'bile salts' 15. The natural occurrence and preparation of allocholic acid. Biochem. J., 85, 236—242.

Anderson, I. G. and Haslewood, G. A. D. (1970) Comparative studies of bile salts. 5α-Chimaerol, a new bile alcohol from the white sucker *Catostomus commersoni* Lácepède. Biochem. J., 116, 581—587.

Anderson, I. G., Haslewood, G. A. D., Oldham, R. S., Amos, B. and Tökés, L. (1974) A more detailed study of bile salt evolution, including techniques for small-scale identification and their application to amphibian biles. Biochem. J., 141, 485—494.

Bergström et al. (1959) and (1960) See References to Chapter 3.

Bogert, C. M. and del Campo, R. M. (1956) The gila monster and its allies. Bull. Am. Mus. Nat. Hist., 109, 1—238.

Brodal, A. and Fänge, R. (1963) Eds. The Biology of Myxine. Universitetsforlaget, Oslo.

Clayton, R. B. (1964) The utilization of sterols by insects. J. Lipid. Res., 5, 3—19.

Danielsson, H. and Einarsson, K. (1964) Further studies on the formation of bile acids in the guinea pig. Bile acids and steroids 141. Acta. Chem. Scand., 18, 732—738.

Deuchar, E. M. (1966) Biochemical Aspects of Amphibian Development. Methuen, London.

Eyssen, H. J., Parmentier, G. G. and Mertens, J. A. (1976) Sulfated bile acids in germ-free and conventional mice. Eur. J. Biochem., 66, 507—514.

Eyssen, H., Smets, L., Parmentier, G. and Janssen, G. (1977) Sex-linked differences in bile acid metabolism of germfree rats. Life Sci., 21, 707—712.

Greenwood, P. H. (1975) A History of Fishes. J. R. Norman. 3rd Edn. Ernest Benn, London.

Greenwood, P. H., Rosen, D. E., Weitzman, S. H. and Myers, G. S. (1966) Phyletic studies of teleostean fishes, with a provisional classification of living forms. Am. Bull. Nat. Hist., 131, 339—456.

Haslewood, G. A. D. (1964) The biological significance of chemical differences in bile salts. Biol. Rev., 39, 537—574.

Haslewood, G. A. D. (1967) Bile Salts. Methuen, London.

Haslewood, G. A. D. (1969) In Bile Salt Metabolism. Eds. L. Schiff, J. B. Carey Jr. and J. M. Dietschy. pp. 155—156. Charles C. Thomas, Springfield, Ill.

Haslewood, G. A. D. (1971) Bile salts of germ-free domestic fowl and pigs. Biochem. J., 123, 15—18.

Haslewood, G. A. D., Ikawa, S., Tökés, L. and Wong, D. (1978) Bile salts of the Green turtle *Chelonia mydas* (L). Biochem. J., 171., 409—412.

Haslewood, G. A. D. and Ogan, A. U. (1959) Comparative studies of 'bile salts' 12. Application to a problem of rodent classification: Bile salts of the cutting-grass, *Thyronomys swinderianus*. Biochem. J., 73, 142—144.

Haslewood, G. A. D. and Tammar, A.R. (1968) Comparative studies of bile salts. Bile salts of sturgeons (Acipenseridae) and of the paddlefish *Polyodon spathula*: a new partial synthesis of 5β-cyprinol. Biochem. J., 108, 236—268.

Haslewood, G. A. D., and Tökés, L. (1972) Comparative studies of bile salts. A new type of bile salt from *Arapaima gigas* (Cuvier) (Family Osteoglossidae). Biochem. J., 126, 1161—1170.

Hofmann, A. F., Mosbach, E. H. and Sweeley, C. C (1969) Bile acid composition of bile from germ-free rabbits. Biochim. Biophys. Acta. 176, 204—207.

Hsia, S. L. (1971) Hyocholic and muricholic acids. In: The Bile Acids. Eds. P. P. Nair and D. Kritchevsky. Vol. 1, pp. 95—120. Plenum Press, New York—London.

Ikawa, S., and Tammar, A. R. (1976) Bile acids of snakes of the subfamily Viperinae and the biosynthesis of C-23 hydroxylated bile acids in liver homogenate fractions from the Adder, *Vipera berus* (Linn). Biochem. J., 153, 343—350.

Kellogg, T. F. (1973) Bile acid metabolism in gnotobiotic animals. In: The Bile Acids. Eds. P. P. Nair and D. Kritchevsky. Vol. 2, pp. 283—304. Plenum Press, New York—London.

Kihara, K., Yasuhara, M., Kuramoto, T. and Hoshita, T. (1977) New bile alcohols 5α- and 5β-dermophols from amphibians. Tetrahedron Letters, 8, 687—690.

Kuramoto, T. and Hoshita, T. (1972) The identification of C_{27}-bile acids in kite bile. J. Biochem. (Tokyo), 72, 199—201.

Kuramoto, T., Kikuchi, H., Sanemori, H. and Hoshita, T. (1973) Bile salts of anura. Chem. Pharm. Bull., 21, 952—959.

Kurozumi, K., Harano, T., Yamasaki, K. and Ayaki, Y. (1973) Studies on bile acids in bear bile. J. Biochem. (Tokyo), 74, 489—495.

Le Gros Clark, W. E. (1962) The Antecedents of Man. 2nd Edn., pp. 288—295. University Press, Edinburgh.

Lindstedt, S. and Sjövall, J. (1957) On the formation of deoxycholic acid from cholic acid in the rabbit. Bile acids and steroids 48. Acta. Chem. Scand., 11, 421—426.

McSherry, C. K., Javitt, N. B., de Carvalho, J. M. and Glenn, F. (1971) Cholesterol gallstones and the chemical composition of bile in baboons. Ann. Surg., 173, 569—577.

Mandava, N., Anderson, J. D., Dutky, S. R. and Thompson, M. J. (1974) Novel occurrence of 5β-cholanic acid in plants: isolation from Jequirity bean seeds (*Abrus precatorius*, L.). Steroids, 23, 357—361.

O'Connell, J. (1970) Absence of sterol biosynthesis in a crab and a barnacle. J. Exp. Mar. Biol. Ecol., 4, 229—237.

Ogan, A. U. (1960) Chemical characters as aids in the study of animal species classification and evolutionary history, with special reference to bile salts. Ph.D. Thesis, University of London.

Patterson, C. (1965) The phylogeny of the chimaeroids. Phil. Trans. Roy. Soc. B, 249, 101—219.

Peric-Golia, L. and Jones, R. S. (1961) Cholesterol-4-C^{14} and bile acids in the guinea pig etc. Proc. Soc. Exp. Biol. Med., 106, 177—180 and 107, 856—858.

Portman, O. W. (1962) Importance of diet, species and intestinal flora in bile acid metabolism. Fed. Proc. 21, 896—902.

Romer, A. S. (1966) Vertebrate Paleontology. 3rd Edn. pp. 98 and 140—141. The University of Chicago Press, Chicago and London.

Samuelsson, B. (1960) Studies on the mechanism of the formation of hyodeoxycholic acid in the pig. Bile acids and steroids 100. Arkiv. für Kemi., 15, 425—432.

Scott, A. M. and Lough, A. K. (1971) The influence of biliary constituents in an acid medium on the micellar solubilization of unesterified fatty acids of the duodenal digesta of sheep. Br. J. Nutr., 25, 307—315.

Sheffer, V. B. (1958) Seals, Sea Lions and Walruses. Stanford University Press, Stanford, California.

Shelton, R. G. J. (1978) On the feeding of the hagfish *Myxine glutinosa* in the North Sea. J. Mar. Biol. Ass. U.K., 58, 81—86.

Simpson, G. G. (1945) The principles of classification and a classification of mammals. Bull. Am. Mus. Nat. Hist., 85.

Sjövall, J. (1959) Dietary glycine and taurine on bile acid conjugation in man. Bile acids and steroids 75. Proc. Soc. Exp. Biol. Med., 100, 676—678.

Smith, A. G. and Goad, L. J. (1975) Sterol biosynthesis in the echinoderm *Asterias rubens*. Biochem. J., 146, 25—33.

Smyth, J. D. and Haslewood, G. A. D. (1963) The biochemistry of bile as a factor in determining host specificity in intestinal parasites, with particular reference to *Echinococcus granulosus*. Ann. N.Y. Acad. Sci., 113, 234—260.

Tammar, A. R. (1970) A comparative study of steroids with special reference to bile salts. Ph.D. Thesis, University of London.

Tammar, A. R. (1974 a, b, c) In: Chemical Zoology. Eds. M. Florkin and B. T. Scheer, a, Bile salts in fishes. Vol. VIII, pp. 595—612; b, Bile salts of Amphibia. Vol. IX, pp. 67—76; c, Bile salts in Reptilia. Vol. IX, pp. 337—351. Academic Press, San Francisco.

Teshima, S-I. and Kanazawa, A. (1974) Biosynthesis of sterols in abalone, *Halotis gurneri* and mussel, *Mytilus edulis*. Comp. Biochem. Biophys., 47, 555—561.

Underwood, G. (1967) A contribution to the classification of snakes. British Museum (Natural History), London.

Van den Oord, A., Danielsson, H. and Ryhage, R. (1965) On the structure of the emulsifiers in gastric juice from the Crab *Cancer pagurus* L. J. Biol. Chem., 240, 2242—2247.

Vonk, H. J. (1962) Emulgators in the digestive fluids of invertebrates. Arch. Internat. Physiol. Biochim., 70, 67—85.

Vonk, H. J. (1969) The properties of some emulsifiers in the digestive fluids of invertebrates. Comp. Biochem. Phsiol., 29, 361—371.

Voogt, P. A. and colleagues (1973, 1974) References in Chemical Abstracts, 78—81.

Wallace, D. G., Maxson, L. R. and Wilson, A. C. (1971) Albumin evolution in frogs. A test of the evolutionary clock hypothesis. Proc. Nat. Acad. Sci. U.S.A. 68, 3127—3129.

Wallace, D. G., King, M-C. and Wilson, A. C. (1973) Albumin differences among ranid frogs: taxonomic and phylogenetic implications. Syst. Zool., 22, 1—13.

Wiggins, H. S. (1955) A study of the occurrence and relationship of the bile acids and their conjugates. Ph.D. Thesis, University of London.

Yamasaki, K., Usui, T., Iwata, T., Nakasone, S., Hozumi, M. and Takatsuki, S-I. (1965) Absence of bile acids in the digestive juice of the swamp crayfish (*Procanbarus clarkii*). Nature, Lond., 205, 1326—1327.

Zandee, D. I. (1966) Metabolism in the crayfish *Astacus astacus* (L.). III. Absence of cholesterol synthesis. Arch. Internat. Physiol. Biochim., 24, 435—441.

CHAPTER 5

The importance to medicine of bile salts

Fortunately for those interested in the subject of this Chapter, an excellent and highly readable monograph by K. W. Heaton (1972) gives the background to our present understanding, the principal factual information and a comprehensive list of references. *Clinics in Gastroenterology*, Vol. 6 (January 1977), gives thorough accounts, by experts, of bile salts in liver and gastro-intestinal disease and in infants and children as well as dealing in more detail with some of the topics briefly covered in this Chapter. Nevertheless, the rate of progress in studies more or less relevant to clinical medicine is so great that any account will be out of date by the time it is published, so that the reader must take this Chapter even more as a bulletin of change than is usual in scientific literature.

5.1. Composition of human bile

Table 5.1 shows the concentration of the chief constituents of gall-bladder bile from some hospital patients without hepatobiliary disease. From the authors' paper, bilirubin is expressed as such (unconjugated) and bile acid as cholic acid; hence the actual concentrations of conjugated bilirubin and bile salt anions will be somewhat greater. Phospholipid was calculated by multiplying the figure for organic phosphorus by 25.

For 27 similar surgical patients, Dam et al. (1971) give, for gall-bladder bile, pH in the range 6.6—8.1 and a mean ± SD total bile acid content of 158.8 ± 8.82 mmol/l. (The molecular weight of cholic acid is 408; hence if this was used as a standard in the analyses,

TABLE 5.1

Gall-bladder bile composition from 20 patients with diseases not affecting the hepatobiliary tract: values in g/dl (From Nakayama and Van der Linden, 1970)

Constituent	Range	Mean
Total solids	3.82–28.60	16.21
Total lipid	3.06–24.05	15.06
Bilirubin	0.07–14.13	0.89
Total bile acid	0.71–9.01	5.87
Phospholipid	0.55–6.75	3.40
Cholesterol	0.14–2.18	1.08
Free fatty acid	0.00–0.38	0.05
Monoglyceride	0.00–7.71	0.49
Diglyceride	0.00–0.09	0.02
Triglyceride	Nil	Nil
Ca^{2+}	0.01–0.05	0.03
Na^+	0.36–0.62	0.47
K^+	0.03–0.09	0.06

158.8 mmol/l corresponds to 6.48 g/dl for comparison with Table 5.1).

Antsaklis et al. (1973) estimated total gall-bladder bile salts by the 3α-hydroxysteroid dehydrogenase method (Chapter 2) in 5 patients without biliary disease and found these to be 83.6 ± 2.7 (mean ± SEM) % of the total lipid in molar terms; they quote similar figures obtained by other workers.

Hepatic bile as obtained at operation or by duodenal aspiration is more dilute than bladder bile. For example, van der Linden and Norman (1967) give averages of 4.50, 0.09, 0.31 and 0.82 g/dl for total solids, bilirubin, cholesterol and phospholipid respectively in hepatic bile collected at operation from 101 patients with gallstone disease: these figures may be compared with those in Table 5.1. A representative composition in health of duodenal bile is shown in Table 5.2. In citing units (wt/vol, or in molar terms), I have no prejudice for or against SI (molar) units, but think it dictatorial, indeed monstrous, that these should have been imposed on those mainly concerned with clinical interpretation without adequate prior consultations and agreement. I also find it difficult to visualize molar amounts without translation into weights. In this book, an attempt is made to be bilingual.

TABLE 5.2
Constituents of duodenal bile from 42 volunteers, aged 18–31. From Dam et al. (1971)

Constituent	Mean ± s.d. of mean
Dry matter	5.5 ± 0.31 g/dl
Total bile acid	38.9 ± 0.32 mmol/l (1.58 ± 0.13 g/dl as cholic acid)
Phospholipid	11.1 ± 1.06 mmol/l lipid-soluble P (approx 0.84 ± 0.08 g/dl as lecithin)
Cholesterol	3.9 ± 0.41 mmol/l (0.15 ± 0.02 g/dl)
G/T ratio (molar ratio: total glycine-conjugated bile acids/total taurine-conjugated bile acids)	1.81 ± 0.15
D/T ratio (molar ratio: total dihydroxy/total trihydroxy bile acids)	1.53 ± 0.09 (T/D = 0.65 ± 0.04)

5.2. Human bile salts (a) in bile

In healthy persons after infancy the principal bile acids are cholic, chenodeoxycholic and deoxycholic, formed as described in Chapter 3; the first two from cholesterol and the third as a secondary product made from cholic acid by microorganisms in the intestine during the enterohepatic circulation (Fig. 1.1). The other likely secondary bile acid, lithocholic acid from chenodeoxycholic acid, is too insoluble even when conjugated to act as a bile salt; it does occur in small amounts and is discussed later. Conjugation is with glycine and taurine; thus there are normally six functioning human bile salts, namely:

Glycocholate (GC)
Taurocholate (TC)
Glycochenodeoxycholate (GCD) } Primary bile salts
Taurochenodeoxycholate (TCD)
Glycodeoxycholate (GD)
Taurodeoxycholate (TD) } Secondary bile salts

Formulae are given in the text: for cholic acid, p. 50; for taurocholate, p. 40; for chenodeoxycholic acid, p. 132; for deoxycholic acid, p. 29 and for conjugates, p. 66.

Human bile also contains a little conjugated trihydroxycoprostanic acid (p. 50), allo(5α)cholic acid (p. 59), lithocholic acid (p. 132) and traces of various other acids including keto acids and ursodeoxycholic acid (UDC, p. 111). Igimi (1976) has reported percentages of UDC of 1.1 to 8.3 (average ± SE, 5.7 ± 0.6) in gall-bladder bile obtained at laparotomy from 10 Japanese patients without hepatobiliary disease; thus UDC may be present in more than traces and this may be relevant to the gallstone problem, discussed below.

The human foetus and infant have a greater capacity for taurine than for glycine conjugation and the adult G/T ratio (Table 5.2) is not usually reached before the age of a few months. This interesting situation should be explored in other vertebrates and, indeed, Subbiah et al. (1977) found that foetal and new-born rabbits had also more taurine conjugates than adults; the bile acid pattern, too, was somewhat different. Little et al. (1975) after a study of bile salt metabolism in near-term foetal Rhesus monkeys suggested that "bile salt metabolic and excretory mechanisms are undergoing developmental maturation at birth". These authors state also that in dogs, foetal mechanisms for bile salt synthesis, conjugation, excretion and turnover "are remarkably mature". Perhaps, therefore, gene expression for these characters is later in some species and especially man. Back and Ross (1973) identified cholic, chenodeoxycholic, deoxycholic, ursodeoxycholic, lithocholic and 3β-hydroxy-5-cholen-24-oic(3β-hydroxycholenic, p. 56) acids in mixed meconium from 11 premature infants born after 32—35 weeks gestation and 13 born at term. Meconium is usually sterile and aerobic enterobacteria were not found in 5 samples in this investigation. Some of the cholic and chenodeoxycholic acid in the mixed meconium was neither conjugated nor sulphated (p. 34) and some of the lithocholic and 3β-hydroxycholenic acid were as sulphate esters only, but all the bile acids also occurred as taurine and glycine conjugates. Back and Ross point out that of course some bile acids and salts (especially deoxycholic acid and its conjugates) almost certainly originated in the mothers but if 3β-hydroxycholenic acid is made in the foetal liver, perhaps the Mitropoulos—Myant pathway for bile acid biosynthesis (Scheme 3.3) may be more favoured in the immature liver, an idea further supported by the finding of 26-hydroxycholesterol (Scheme 3.3) in relatively large amounts in meconium. Délèze et al. (1977) have found that 18 specimens of amniotic fluid obtained by amniocentesis from women without disease all contained 3β-hydroxycholenic acid as well as cholic and chenodeoxycholic acids. Délèze et al. suggest that the bile acid pattern in amniotic fluid might provide an index of foetal liver maturity.

Dam et al. (1971) found a G/T ratio (mean ± SD of mean) of 2.44 ± 0.28 (range 0.99—5.19) for gall-bladder bile in the 27 surgical patients mentioned above and 1.81 ± 0.15 (range 0.40—4.94) for duodenal bile (Table 5.2). For the T/D ratio (Table 5.2), investigators have given values in the range 0.66—1.20 for bladder bile in health (Nakayama and van der Linden, 1970); Dam et al. (1971) give 0.57 ± 0.06 (mean ± SD of mean) for their 27 surgical cases and 0.65 ± 0.04 (range 0.29—1.31) for duodenal bile (Table 5.2).

5.3. Human bile salts (b) in normal blood

In health, the concentration of circulating bile salts is very low. Analyses are usually made on serum, or occasionally plasma; there does not seem to be reliable information as to distribution between plasma and red cells. Some values taken from recent publications by workers using what appear to be well-tested methods are given in Table 5.3.

Enzyme methods use the 3α-hydroxysteroid dehydrogenase mentioned on p. 38; 7α-hydroxysteroid dehydrogenase could be used for differential analysis. Gas-liquid chromatography of course does permit differential measurement and if hydrolysis is done with mild (enzymic) methods (p. 38) the sometimes substantial losses incurred in alkaline hydrolysis of very small quantities of conjugates can be avoided.

Some clinicians regard the T/D ratio as having diagnostic significance, as mentioned later, and so differential analysis may be important. By their radioimmunoassay (RI) method, Demers and Hepner (1976) claim to be able to estimate separately the glycine conjugates of the three chief bile acids and also of glycolithocholic acid sulphate. It is fair to say, however, that RI is still in its infancy and improvements are to be expected.

Hofmann et al. (1974) showed that their RI method could demonstrate a rise in serum cholic acid conjugates after a liquid test meal in normal subjects. Fasting values were less than 1.0μmol/l and rose to about $1.5—2.7 \mu$mol/l after the meal thrice in a 24-h period. The same group (La Russo et al., 1975) using RI, have developed an intravenous cholylglycine (glycocholate) tolerance test which they claim is of more sensitive diagnostic value in liver disease than either transaminase measurements or the bromsulphthalein retention test as usually carried out.

TABLE 5.3
Serum bile acid or bile salt concentrations in normal subjects

Method	Bile acids or bile salts (μmol/l, total except where otherwise stated)	Reference
Gas-liquid chromatography (GLC)	0.6–4.6	
Thin-layer chromatography/fluorimetry	2.2–4.2	
Enzyme/fluorimetry	0–4.7 (males)	Schwartz et al. (1974) who give references to the above results.
	1.0–8.2 (females)	
Enzyme/fluorimetry	0.3–8.3 (males)	
	3.6–9.3 (females)	Fausa, 1976
Enzyme/fluorimetry	2.5 ± 1.4	Mashige et al., 1976
Enzyme/fluorimetry[a]	3.6–12.6	Henegouwen et al., 1974
GLC	0.73–6.13 (as cholic acid)	Henegouwen et al., 1974
	trace–2.4 (cholic acid)	Henegouwen et al., 1974
	trace–2.6 (chenodeoxycholic acid)	Henegouwen et al., 1974
	trace–2.6 (deoxycholic acid)	
Radioimmunoassay	≥1.0 (cholic acid conjugates)	Korman et al., 1975
Radioimmunoassay	0.55–1.8 (cholic acid conjugates)	Murphy et al., 1974
	0.1–0.85 (glycocholate)	Demers and Hepner, 1976
	0.08–0.70 (glycochenodeoxy-cholate)	Demers and Hepner, 1976
	0.01–0.09 (glycodeoxycholate)	Demers and Hepner, 1976
Radioimmunoassay	0.18–1.25 (cholic acid conjugates)	Matern et al., 1976

[a]This method involves the transfer of 2H from the 3α-OH of bile acids to resazurin to give a fluorophore, increasing the sensitivity.

5.4. Human bile salts (c) in normal urine

Until recently, it was supposed that little or no bile salts occurred in urine in health, and even in obstructive jaundice detection proved difficult. However it is now realised that sulphate and glucuronate esters may be present as well as taurine and glycine conjugates and when methods for cleaving these are applied, bile acids can readily be detected. J. Sjövall and his colleagues have devised methods for (1) concentration of the urinary bile salts by absorption on to an Amberlite resin, followed by elution from the resin, (2) separation of the eluted bile salts on a lipophilic anion-exchanger into (a) unconjugated bile acid ions, (b) conjugates with glycine and taurine, (c) sulphate ester ions and (d) conjugated sulphate ester ions, and (3) solvolysis and estimation of the various fractions (a)—(d) by computerized GLC—mass spectrometry. This methodology is capable of the most detailed analysis yet described and in an important paper Almé et al. (1977) estimate that $6.4-11\mu$ moles of total bile acids were excreted daily in urine from 5 healthy persons. About 30 bile acids were wholly or partially identified and the authors conclude "The physiological or clinical significance of the complexity of bile acids in urine is not clear. The bile acid metabolite profile in urine is the net result of metabolic reactions and transport processes in the liver, intestinal tract and kidney".

5.5. Human bile salts (d) in faeces

Most, if not all, faecal bile acids are unconjugated and a very complex mixture of substances can be extracted. The method of Grundy et al. (1965) is still in use for total and differential analysis of faecal bile acids. In this procedure, homogenized faeces are given a mild alkaline saponification in aqueous ethanol, after which neutral lipid (sterols etc.) is extracted with light petroleum and can be estimated separately. The alkaline ethanolic portion is then saponified more strongly and, after acidification, acids are extracted with chloroform—methanol. Purification on a fluorosil column may or may not be required at this stage, which is followed by methylation, separation by thin-layer chromatography and quantitation by GLC. GLC—mass spectral computerized analysis is an obvious possible refinement here. It is certainly unsafe to use 3α- or 7α-hydroxysteroid enzymes for assay, for

OH groups at C-3 or C-7 could be inverted, converted to keto groups or removed: however, an approximate analysis (perhaps accounting for 90% of faecal bile acids) is possible by such a method and a recent example (Sheltawy and Losowsky, 1975) gives a total output by 11 subjects without gastrointestinal or liver disease of 0.10—0.36 g (mean, 0.20 g) in 24 h, in fair agreement with GLC methods. Eneroth and Sjövall (1971) describe other methods for faecal bile acids, including one of their own. Carey (1973) listed 25 bile acids detected in human faeces: deoxycholic and lithocholic acids accounted for about half the weight. Campbell and McIvor (1975), who use a simplified Grundy analysis, report a total 24-h excretion of 0.37 ± 0.13 g (range 0.115—0.685 g) of bile acids by 26 "normal" subjects.

5.6. The solubility of cholesterol in bile

The means by which cholesterol is held in solution in bile is of great medical interest for two reasons (1) because excretion via the bile either as cholesterol itself or after conversion to bile salts is the only way in which cholesterol can leave the body to any quantitatively significant extent and (2) because cholesterol-containing gallstones are a common cause of disease.

Loss of cholesterol from the body may lower the plasma cholesterol and thus may affect formation of atheromatous deposits in the arteries. Except in persons with greatly raised concentrations of plasma cholesterol (e.g. in "idiopathic" familial hypercholesterolaemia), who have a much increased chance of atherosclerotic crises (coronary occlusion or cerebral thrombosis), there is some controversy about the relation between blood cholesterol concentrations and atheromatous disease. However, it is generally conceded that lowering blood cholesterol is beneficial so that there is good reason to pay attention to its loss via the bile.

There is at present much practical interest in the possible dissolution in situ of gallstones by giving bile acids and this treatment, too, is rationally based on understanding biliary cholesterol solubilization.

I have argued previously that man is often eating like a carnivore with the physiology of an omnivore and that there has not yet been time for selective adaptation to cope with the bilirubin and cholesterol that must be excreted via the bile on such a diet (Haslewood, 1967).

Human medical skill is likely to prevent such adaptation and people generally wish to solve without serious inconvenience what may be essentially a social problem. Such a view of course implies that a large part of the biliary bilirubin and cholesterol is exogenous. As to the bilirubin, human bile is very much darker than that of other primates I have seen: I know of no reports on the bilirubin excretion of true vegetarians. It is conceded that a part of the cholesterol in bile is exogenous, although the proportion is uncertain and variable.

The above considerations do not, and should not, hinder the progress of studies on the formation of gallstones and it has been clear for some time that supersaturation with cholesterol is a common condition of human bile, in contrast to that, for example, of pigs, sheep, oxen and dogs (Holzbach et al., 1973).

D. M. Small and his colleagues gave quantitative significance to the idea that the principal substances affecting cholesterol solubility in bile are lecithin and bile salts and after a series of careful studies with cholesterol, mixtures of bile salts such as normally occur in human bile and lecithin from egg-yolk, they could express their results by a phase diagram such as Fig. 5.1.

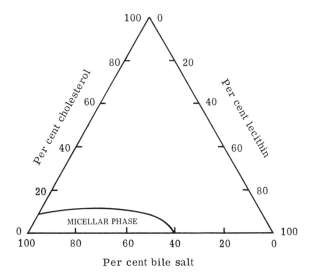

Fig. 5.1. Phase diagram showing the area of micellar solubility of cholesterol in mixtures of lecithin and bile salts, both from human bile.

Within the "micellar phase" area shown on this diagram, that is within the molar percentages bounded by this area, cholesterol is held in solution as micelles with lecithin and bile salts: outside this zone cholesterol is not soluble. Thus, if the three molar percent concentrations are determined for any bile, the point resulting will show whether or not the bile is supersaturated with cholesterol. Bile so supersaturated is sometimes described as "lithogenic".

The micellar-phase area originally proposed by Small and his colleagues (e.g. Admirand and Small, 1968) was somewhat larger than that shown in Fig. 5.1, mainly because of experimental difficulties discussed in detail by Carey and Small (1974). Other workers, for example Hegardt and Dam (1971) and Holzbach et al. (1973) have proposed smaller zones: Fig. 5.1 shows approximately an agreed area.

J. A. Lucy and his colleagues, using negative-staining electron microscopy, have observed cigar-shaped macromolecular assemblies both in natural bile and in cholesterol—lecithin—bile salt mixtures: these particles may be composed of disc-shaped mixed micelles intercalated with layers of water (Howell et al., 1970). It is generally supposed that lecithin and bile salts are virtually the only substances that determine cholesterol solubility in bile; this assumption is supported by the common experience that biles whose cholesterol/bile salt lecithin proportions fall outside the micellar zone in Fig. 5.1 are often found in patients with gallstones. Nevertheless the correlation is far from complete (Holzbach et al., 1973) and one wonders whether the mucin or some other unexplored constituent of bile might play a more important role than so far recognised acting, for example, as protective colloids and hindering crystallization of cholesterol. Cholesterol "supersaturation" is common in man and baboons, but frequently does not result in gallstones.

5.7. The gallstone problem

Except in man and perhaps baboons (McSherry et al., 1971) gallstones are very uncommon in wild animals and those kept in the usual conditions of domesticity. Cattle stones seem to be mainly composed of bilirubin, I have seen one such concretion from a sheep and Wood et al. (1974) found many in foetal sheep gallbladders. The rare pig gallstones consist of calcium salts of lithocholic acid and $3\beta,6\alpha$-dihydroxy-5β-cholanic acid (Haslewood, 1967). Gallstones of various compositions have been induced, for example, in dogs, cats, rats, mice, rabbits, guinea pigs, ground squirrels, prairie dogs and hamsters (Schoenfield and Sjövall, 1966; Freston and Bouchier, 1968; Sutor et al., 1969; Englert et al., 1969; Holzbach et al., 1976).

Some human populations are particularly prone to gallstones, for example American Indians (Adler et al., 1974), whereas others (e.g. Africans; Heaton et al., 1977) are almost without them. There is a relationship between the prevalence of cholesterol stones and the composition of bile in human races (Oviedo et al., 1977). Stones consist usually mainly of cholesterol and/or calcium bilirubinate, carbonate, palmitate or phosphate with an organic matrix; the older classification into cholesterol, pigment or mixed stones is outdated (Sutor and Wooley, 1971, 1973, 1974). Studies by physical methods of the actual crystalline and macromolecular assemblies that exist in gallstones may throw light on the physiology of their formation (e.g. Bills and Lewis, 1975).

Stones not consisting mainly of cholesterol cannot be removed except by surgery, but in recent years there has been considerable progress towards a successful "drug" treatment of radiolucent cholesterol stones (i.e. those containing little or no calcium salts), particularly if these are small and the patient has a functioning gallbladder. Thistle and Schoenfield (1969) reported that the "lithogenic potential" of patients' bile was decreased by giving chenodeoxycholic acid (3.16) and Danzinger et al. (1972) and later Bell et al. (1972) that cholesterol gallstones could be dissolved in situ by this treatment. This discovery caused a considerable sensation, for oral therapy had not previously succeeded and there are many patients for whom surgery is inadvisable or unacceptable.

An immediate biochemical worry was that the same intestinal microorganisms that convert cholic to deoxycholic acid also make lithocholic acid (3.15) from chenodeoxycholic acid, as shown in Scheme 5.1.

Now, it is known from earlier work that lithocholic acid is toxic to the liver in many species and pyrogenic in man, so that the prospect of greatly increased lithocholate production as well as the possibility of chenodeoxycholic acid toxicity itself required careful study. The position was reviewed in detail by Palmer (1972) and later work showed that chenodeoxycholic acid was indeed hepatotoxic to Rhesus monkeys (*Macaca mulatta*: Heywood et al., 1973; Webster et al., 1975; Dyrszka et al., 1976) baboons (*Papio anubis* and *P. cyanocephalus*: Morrissey et al., 1975) and rabbits (Fischer et al., 1974).

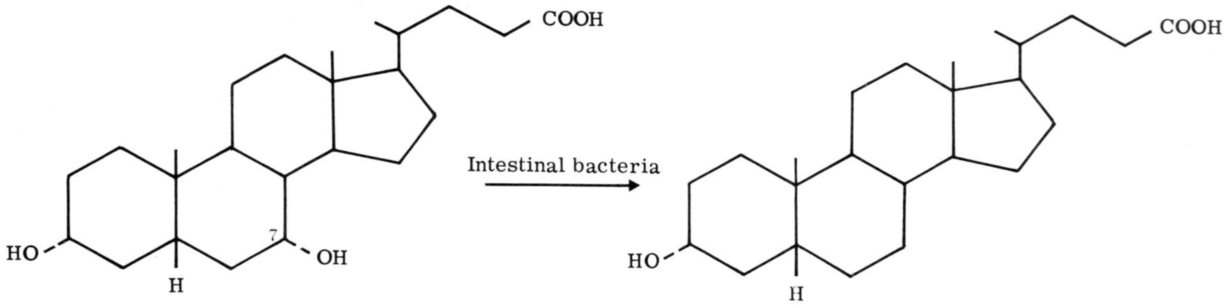

3.16 Chenodeoxycholic acid   **3.15** Lithocholic acid

Scheme 5.1. Conversion of chenodeoxycholic acid to lithocholic acid.

Palmer originally showed that man could sulphate lithocholate quite efficiently and Schwenk et al. (1978) have demonstrated that our nearest relative, the chimpanzee, has a quantitatively similar ability. This is not true of some more remote primates, for example Rhesus monkeys. After a thorough study, Cowen et al. (1975a) suggest that sulphation may enhance the faecal excretion of lithocholic acid and its conjugates and may "be a biotransformation of considerable protective advantage", especially in man. Such a mechanism might account for the relative lack of toxicity of chenodeoxycholic acid to our species.

Iser et al. (1975) were able "to confirm that in suitable patients, chenodeoxycholic acid given by mouth effectively dissolves gallstones and that in doses up to 1.5 g per day it appears safe". A transient rise in serum transaminase may occur and also diarrhoea in some patients, but no serious complications have been reported. Dowling (1976) recommended a dosage of 15 mg of chenodeoxycholic acid per kg of body weight per day and stated that those most likely to respond are non-obese patients with small radiolucent stones in functioning gall-bladders, in whom the overall efficacy (dissolution of stones) is 80—90%. On the same occasion, Schoenfield et al. (1976) said that "the medical dissolution of cholesterol gallstones by administration of chenodeoxycholic acid is still experimental therapy" and that "other agents, adjuvents or dietary manipulations should be sought for safe and more rapid and effective dissolution of gallstones".

At the time of writing, a clinical trial to involve about 1000 patients is in progress in the U.S.A. and a number of centres elsewhere (for example, Pedersen and Bremmelgaard, 1976) are doing similar investigations, which usually involve also a study of the effects of the treatment on constituents of bile obtained by duodenal intubation.

Chenodeoxycholic acid therapy appears to exert its in vivo solubilising effect chiefly by converting the circulating bile salt pool almost completely to chenodeoxycholic acid conjugates and also by decreasing the concentration of cholesterol in the bile (Northfield et al., 1975), possibly by inhibition of the HMG-CoA reductase enzyme that may be rate-controlling for cholesterol biosynthesis (p. 68). Decreasing biliary cholesterol leads to a less saturated bile, a situation expressed quantitatively by a smaller "lithogenic" or "saturation index". This index is defined as the ratio of the molar concentration of cholesterol actually present to the maximal amount that would dissolve at the phospholipid (lecithin)-bile salt concentrations of the sample. Thomas and Hofmann (1973) suggested a simple method of calculating this, using either the original curve of Admirand and Small (1968) or that of Hegardt and Dam (1971) and Holzbach et al. (1973), as depicted in Fig. 5.1. For this latter line marking the limit of maximum cholesterol solubility, the equation for the plot molar % cholesterol versus molar ratio phospholipid/phospholipid + bile salts is:

(1) $y = 3.082 - 0.804x + 117.05x^2 - 204.94x^3$
(up to $x = 0.370$)

where y = molar % cholesterol at saturation and x = the molar ratio phospholipid/phospholipid + bile salts. For higher values of x, where the maximum solubility boundary approaches that of Admirand and Small, the equation:

(2) $y = 117.4 - 755.0x + 1.756x^2 - 1.377x^3$

may be used.

Equations (1) or (2) enable one to calculate the concentration of cholesterol at saturation (y) for any values of phospholipid and bile salts and then the lithogenic index is the ratio: mole % of cholesterol in the sample / y.

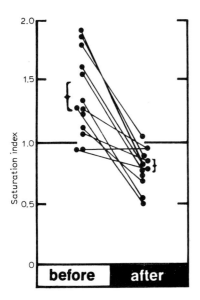

Fig. 5.2. The effect of chenodeoxycholic acid therapy on the extent of saturation of bile with cholesterol.

It is obvious that at saturation the index = 1.0 and values greater than this indicate a bile supersaturated with cholesterol. Mok et al. (1974) prefer the term "saturation index", since "normal individuals often have supersaturated bile but do not form gallstones". The effect of chenodeoxycholic acid therapy in lowering the index is frequently expressed as in Fig. 5.2, taken from Iser et al. (1975).

When chenodeoxycholic acid therapy is stopped the saturation index of bile rises and stones often reform (Dowling, 1976).

Toouli et al. (1975) reported that the treatment of 9 patients with radiolucent stones in the gall-bladder or biliary tree with a mixture of cholic acid (750 mg) and soya-bean lecithin (250 mg) thrice daily for 6 months resulted in the disappearance of stones in 2 cases. Clearly this regimen is less effective than chenodeoxycholic acid therapy, but it did result in a marked lowering of the biliary lithogenic index. Pomare et al. (1976) found a similar effect after feeding wheat bran (at least 20 g daily for 4—6 weeks) to 10 patients with functioning gall-bladders containing radiolucent stones; this regimen also increased the synthesis of chenodeoxycholic acid and the size of its circulating pool in 6 symptomless subjects.

Quite recently it has been suggested that ursodeoxycholic acid, the 7β-epimer of chenodeoxycholic acid (Formula 4.17), might be even more effective and less potentially toxic than chenodeoxycholic acid itself (Makino and Nakagawa, 1976). Ursodeoxycholic acid may comprise more than 8% of the total bile acids in gall-bladder bile (Igimi, 1976): it is biochemically linked with chenodeoxycholic acid via 3α-hydroxy-7-oxo-5β-cholanic acid (p. 70) and by this route could also give rise to some lithocholic acid in an intact animal. Its metabolism and therapeutic value are certain to be thoroughly tested (e.g. Igimi et al., 1977; Fedorowski et al., 1977; Maton et al., 1977; Nakagawa et al., 1977).

Summarising, we can say that an oral therapy for dissolving radiolucent cholesterol gallstones is available and likely to be improved. Several well-equipped centres are engaged in elucidating the rational basis of this treatment. Marks et al. (1976) give a thoughtful and detailed review of the problem.

5.8. The bile salt "pool"

Vlahcevic et al. (1970) found by the isotope dilution method that 8 patients with gallstones had an average of 1.29 g of total bile acids circulating as their conjugates in the enterohepatic "pool"; the corresponding figure for 9 patients without gallstones was 2.38 g. It seemed that the gallstone patients produced less cholic acid (206 mg/day) than those without stones, who made 358 mg of cholic acid per day. However, Mok and Dowling (1974) question the validity of both the isotope dilution and isotope washout methods of measuring bile salt pools. The latter method gave results closely corresponding with daily secretion rates in Rhesus monkeys with permanent biliary fistulae. In patients without gallstones, the pool size was 2.0—2.5 g as measured by the isotope dilution method, which is the only one applicable to man. The average fractional turnover was about 0.3, i.e. about 600—800 mg/day, which is greater than measured faecal excretion of bile acids (Section 5.5). The discrepancy is unexplained but perhaps the isotope dilution method measures more bile salts than those actually in the enterohepatic circulation.

5.9. Bile salts in liver disease

It has become clear that in liver disease where cholestasis occurs there are quite profound alterations both in bile pigments and bile salts. The position for bilirubin conjugation was mentioned on p. 8 and there have been several reports of the production of unusual bile acids. For example, Summerfield et al. (1976a) mention, with references, that cholic, chenodeoxycholic, deoxycholic, ursodeoxycholic, lithocholic, 3α-hydroxy-5α-cholanic, 3β-hydroxy-5-cholen-24-oic acids and an unsaturated monohydroxy acid have been found in blood plasma and urine in this type of disease: they themselves also identified $3\beta,7\alpha$-dihydroxy-5β-cholanic, hyodeoxycholic, hyocholic and $3\alpha,7\beta,12\alpha$-trihydroxy-5β-cholanic acids as well as 26-hydroxycholesterol and (probably) 5β-cholestane-$3\alpha,7\alpha,26$-triol. In addition, a high percentage of the urinary bile acids may be excreted as sulphate esters, possibly to some extent made in the kidney (Summerfield et al., 1976b, 1977; Almé et al., 1977). In some patients with hepatobiliary disease, $3\alpha,7\alpha$-dihydroxy-12-oxo-5β-cholanic acid has been found in fistula bile; its metabolic origin is unclear (Kikuchi, 1972). This variety of substances recalls the remarkable bile acids produced by surgically jaundiced rats and mice in the experiments of E. A. Doisy and his colleagues in the 1960's (references in Haslewood, 1964).

In jaundice, blood bile salt concentrations are of course greatly increased. Murphy (1972) by enzyme-fluorimetry, found total serum bile acids between 51 and 185 µmol/l in patients with raised serum bilirubin and alkaline phosphatase. Javitt et al. (1976), by GLC, found up to 14 mg/100 ml (343 µmol/l as cholic acid) in serum from children with familial cholestasis and Demers and Hepner (1976), by radioimmunoassay, detected, elevated levels of glycine-conjugated cholic and chenodeoxycholic acid respectively in serum from 22 cholestatic patients; however, they found higher value in cirrhosis, hepatitis and liver cancer. Bron et al. (1977) measured bile acid concentrations in serum, cerebrospinal fluid and brains of patients with fulminant hepatic failure. Bile salt concentration in serum may be raised when bilirubin is not (e.g. Fausa and Gjone, 1976). It is generally agreed that some of the bile salts circulating in the blood in cholestatic disease are sulphated (e.g. Stiehl et al., 1975; Makino et al., 1975) and bile salt glucuronides have been found in urine and plasma in those conditions (Back, 1976; Frohling and Stiehl, 1976).

In spite of the traditional belief that the sinking of flowers of sulphur in jaundiced urine (Hay's test) is caused by the surface tension-lowering effect of bile salts, it was not until 1968 that these compounds were actually proved to be present in urine from a patient with obstructive jaundice. Even so, Gregg (1967) found only about 10 mg each of cholic and chenodeoxycholic acids as their conjugates in a 24-h urine collection. Taking sulphated bile acids into account, Back (1973a) found up to 85 mg/24h in patients with obstructive jaundice. This quantity of human bile salts in a litre of urine is far below the amount necessary to cause powdered sulphur to sink and it is likely that the rarely-seen positive responses to Hay's test have some other cause. Stiehl (1974) found a maximum total daily urinary excretion of about 88 mg of bile salts in patients with obstructive jaundice; more than half the bile salts were sulphated. Also in obstructive jaundice, Makino et al. (1975) found up to about 24 mg of bile salts in a daily urine. Larger amounts were excreted in acute hepatitis and a high proportion of the bile salts were sulphates. In cholestasis, as in other conditions involving increased concentrations of bile salts in blood, pruritus (itching) is common. It can often, but not always, be relieved by eating an anion-exchange resin, for example cholestyramine, that by simple opposite charge attraction "sequesters" intestinal bile salt anions, so breaking the enterohepatic circulation and causing a much increased loss of bile salts in the faeces. Although this treatment stimulates biosynthesis of fresh bile salts; it does usually lower the plasma concentration and may relieve itching.

W. N. Mann (1974), in a thoughtful review of biliary cirrhosis, said that "cholestasis means no more than a failure of the bile flow through the biliary channels". Hans Popper and his colleagues believe that some of the effects of cholestasis are caused by injury to the endoplasmic reticulum of the hepatocytes, brought about by the detergent action of accumulated bile salts (Schaffner et al., 1971).

The late James B. Carey, Jr., a much respected pioneer of bile salt studies in human disease who devised perhaps the first reasonably accurate method for measuring serum bile salt concentrations, attached prognostic significance to the serum T/D ratio (Table 5.2). He said that "with severe liver injury a disproportionate increase in the dihydroxy bile acid concentration reduced the trihydroxy dihydroxy ratio to less than 1.0. Persistence of this reversed ratio is frequently followed by coma, death or both" (Carey, 1958). Later work with present-day methods has confirmed that a low T/D does indeed often occur in liver disease (e.g. Murphy, 1972) but its prognostic value is less clear (see also Heaton, 1972). Murphy (1972) also agreed with others in finding a low G/T ratio (Table 5.2) in cholestatic liver disease and this may occur without cholestasis. The biochemical basis of such alterations in conjugating capacity of diseased human liver has long been studied, for example by Scherstén (1967).

In cirrhosis, there may be little or no deoxycholic acid or its conjugates in bile and plasma and this may be because conditions in the intestine then become unfavourable for the action of 7α-deoxygenating bacteria (Yoshida et al., 1975).

As mentioned in Section 5.3, La Russo et al. (1975) have proposed an intravenous glycocholate tolerance test that may be of particular diagnostic value in chronic hepatitis; it always gave abnormal results in jaundiced patients with hepatic or biliary disease. After giving 0.5 g of cholate by mouth, Matern et al. (1977) were able to distinguish the "tolerance" (rate of fall of serum cholate, measured by RI) of patients with liver cirrhosis from that of those with additional portal hypertension. Both groups had abnormally increased tolerance. Thjodleifsson et al. (1977) have assessed the value of the rate of disappearance of cholyl-^{14}C-1-glycine (glycocholate labelled in the glycine) from blood plasma as a test of hepatocellular disease.

5.10. Bile salts in biliary atresia

This infantile condition, which may prove fatal or be "outgrown" in later life, is an absence or gross deficiency of hepatic ductular function, possibly caused by developmental or acquired metabolic defects. It may amount to a virtual absence of extrahepatic ducts (Javitt, 1974).

In the duodenal fluid from two unrelated children with intrahepatic bile duct anomalies Eyssen et al (1972) found large quantities of the C_{27} trihydroxycoprostanic acid. This substance is a biochemical precursor of cholic acid in man (p. 50) and Hanson et al. (1975) suggested that in such patients the metabolic defect is the inability to hydroxylate trihydroxycoprostanic acid at C-24 and so to remove the terminal three carbon atoms to form cholic acid. The inheritance seemed to be via an autosomal recessive gene.

Makino et al. (1971) found 3β-hydroxy-5-cholen-24-oic acid and allo(5α)lithocholic acid as well as 24-hydroxycholesterol in urine from 4 children with extrahepatic biliary atresia; lithocholic acid was also present in one case. Cholic and chenodeoxycholic acid conjugates, too, are found in urine from these patients and the bile salts are partially sulphated. In a similar case, Back (1973b) also detected ursodeoxycholic acid. Makino et al. (1971) gave ^{14}C-4-cholesterol to one of their patients and could detect less than 5% of the total bile acid radioactivity in their combined monohydroxy bile acid —dihydroxysterol fraction. They suggested that cholesterol may not be a direct precursor of these substances. Back (1973b) on the other hand, thought it possible "that the different atypical primary bile acids found in extrahepatic atresia result from the metabolism of a common precursor, synthesized by the mitochondrial oxidation of cholesterol, possibly 3β-hydroxy-5-cholenoic acid or 3-oxo-4-cholenoic acid".

5.11. Other diseases involving bile salts

Weber et al. (1973) found that children with cystic fibrosis excreted about seven times as much bile acid as normal children. Their output was comparable to that of patients with ileal resection, in whom, of course, the chief site for bile salt absorption has been removed (p. 3).

It has been suggested that an ileal defect might account for the failure to absorb bile salts in cystic fibrosis and indeed that an intestinal fault other than pancreatic insufficiency could contribute to malabsorption in these patients. However, the condition remains largely unexplained.

In untreated coeliac disease, the gallbladder is somewhat inert and the enterohepatic circulation of bile salts sluggish. The total bile salt pool may be more than 9 g, as compared with a normal pool of about 3 g (Low-Beer et al., 1973).

In Wilson's disease, Cowen et al. (1975b) found normal biliary bile acids and a decreased glycocholate tolerance.

As mentioned on p. 53, patients with the rare disease cerebrotendinous xanthomatosis may have in their bile a number of 5β-cholestanes hydroxylated at C-3α,C-7α,C-12α,C-25 and elsewhere in the side-chain (Table 2.1). The disease is characterised inter alia by defective cholesterol and bile acid metabolism; sufferers are probably homozygous for an autosomal recessive gene. The presence of considerable amounts of 5α-cholestan-3β-ol in the nerve tissue as well as the strange bile alcohols makes this condition one of great biochemical interest; it is described in detail by Salen and Mosbach (1976).

Dowling and his colleagues have described a case of congenital bile acid deficiency. The patient, a woman of 28, has had lifelong intractable and severe constipation and steathorrhea, partially corrected by taking a commercial bile salt preparation. The bile salt pool size and daily rate of synthesis are much less than normal. At the time of writing, analysis of the bile is incomplete (Iser et al., 1977).

In diverticulosis of the small bowel in elderly women, concretions (enteroliths) may form especially in the jejunum. These consist almost entirely of choleic acids, i.e. co-ordination compounds of deoxycholic acid with fatty acids in molar ratio 8/1, formed round an insoluble (in my experience, vegetable) residue. I have found no consistent pattern in the fatty acids involved. Bradbrook and Bird (1977) describe a recent case.

There is one recorded example of a cholic-chenodeoxycholic acid stone in the afferent duojejunal loop of a man of 52 who had had a gastrojejunostomy since the age of 4 (Fisher et al., 1965).

In these cases, presumably, prolonged contact with intestinal bacteria causes hydrolysis of conjugates, and in diverticulosis the formation of deoxycholic acid from cholic acid, the former then precipitating gradually as a choleic acid in the presence of fatty acids at the prevailing pH. In a recent case I found that the "fat" (light petroleum-soluble) material from the enterolith consisted of a complex mixture mainly of long-chain fatty acids in the acid (protonated) form, i.e. not as soaps, showing that the ambient pH in life must have been quite low.

Patients with the stagnant loop syndrome suffer from malabsorption and steathorrhea. Northfield et al. (1973) have found that bile acid analyses of jejunal samples, obtained by intubation in these patients, had considerable diagnostic value and were "sufficiently simple and rapid for use as a routine diagnostic procedure in a laboratory not having any special experience in bile acid chemisty". Northfield (1973) has suggested that steathorrea in this condition might be caused by intraluminal precipitation of bile acids liberated by bacteria from their conjugates in increased amounts.

Failure of bile salts to reach their intestine means that the micellar mechanism for fat absorption (p. 12) cannot work or works imperfectly. Nevertheless, absorption of triglycerides can apparently be quite effective, probably because processes other than those described on p. 12 can then play a greater part. What these other mechanisms may be is not clear; they may involve absorption of fatty acids directly into the portal circulation, as normally occurs with short-chain acids. That implies that some hydrolysis of triglycerides can take place in the absence of bile salts and this does indeed seem to be the case. Neutral fat (i.e. fat that cannot be hydrolysed) is poorly absorbed in bile salt deficiency and it may be desirable to supply extra fat-soluble vitamins, such as A and K.

5.12. The breath test

Fromm and Hofmann (1971) devised an ingenious test for bacterial overgrowth in the intestine. A 10μCi dose of glycocholate, labelled with ^{14}C in the glycine portion (cholic acid conjugated with ^{14}C-1-glycine) is given orally and CO_2 from breath samples is trapped and counted for ^{14}C-radioactivity at 2-h intervals. Many types of intestinal bacteria remove the glycine and catabolize it, giving $^{14}CO_2$, the production of which is thus a crude measure of total bacterial population. This breath test is very easy to do and has proved of considerable clinical value. If combined with an assay of ^{14}C excreted in the faeces it is capable of distinguishing between overgrowth of deconjugating bacteria in the small intestine (contaminated small bowel syndrome) and in the colon (bile acid malabsorption syndrome); it seems to be the most sensitive test available for ileal disease (Thaysen and Pedersen, 1977). Its value in Crohn's disease and other conditions has been examined by Scarpello and Sladen (1977).

In healthy persons, the presence of unconjugated bile acids in the lower jejunum is quite common and in the lower ileum there may be a 1.0 mmolar concentration of these substances (Northfield and McColl, 1973).

5.13. Effect of diet on cholesterol balance

As mentioned in Section 5.6, an important reason for studying cholesterol in bile is that biliary excretion, either as cholesterol itself or as bile salts, is the only important method by which this sterol can leave the human body. In his book Heaton (1972) has a thoughtful and scholarly chapter on the effect of diet on biliary output and blood plasma concentrations of cholesterol. It seemed at one time that dietary fibre containing lignin, which is known to bind to bile salts in vitro, might have a beneficial effect by acting like cholestyramine and so removing cholesterol (after conversion into bile salts) from the enterohepatic circulation.

After feeding raw wheat bran to healthy people (see Section 5.7) Pomare et al. (1976) concluded that "binding of bile salts by bran ... either does not occur in vivo or is limited to the colon". Kritchevsky and Story (1976) give a discussion on the effects of dietary fibre on lipid and bile salt metabolism.

Other dietary regimens have given confusing and inconclusive effects on biliary output of bile salts and cholesterol; however, Heaton (1972) concludes that "there is much evidence that diets containing refined carbohydrate have adverse effects on cholesterol-bile acid metabolism". He speculates that a rapid rise in blood sugar and thus in blood insulin "might lead to alteration in hepatic cholesterol metabolism".

References

Adler, R. D., Metzger, A. L. and Grundy, S. M. (1974) Biliary lipid secretion before and after cholecystectomy in American Indians with cholesterol gallstones. Gastroenterology, 66, 1212—1217.

Admirand, W. H. and Small, D. M. (1968) The physicochemical basis of cholesterol gallstone formation in man. J. Clin. Invest., 47, 1043—1052.

Almé, B., Bremmelgaard, A., Sjövall, J. and Thomassen, P. (1977) Analysis of metabolic profiles of bile acids in urine using a lipophilic anion exchanger and computerized gas-liquid chromatography-mass spectrometry. J. Lipid Res., 18, 339—362.

Antsaklis, G., Lewin, M. R., Sutor, J. D., Cowie, A. G. A. and Clark, C. G. (1973) Gallbladder function, cholesterol stone and bile composition. Gut, 16, 937—942.

Back, P. (1973a) Identification and quantitative determination of urinary bile acids excreted in cholestasis. Clin. Chim. Acta, 44, 199—207.

Back, P. (1973b) Synthesis and excretion of bile acids in a case of extrahepatic biliary atresia. Helv. Med. Acta, 37, 193—200.

Back, P. (1976) Isolation and identification of a chenodeoxycholic acid glucuronide from human plasma in intrahepatic cholestasis. Hoppe—Seyler's Z., 357, 213—217.

Back, P. and Ross, K. (1973) Identification of 3β-hydroxy-5-cholenoic acid in human meconium. Hoppe—Seyler's Z., 354, 83—89.

Bell, G. D., Whitney, B. and Dowling, R. H. (1972) Gallstone dissolution in man using chenodeoxycholic acid. Lancet, 2, 1213—1216.

Bills, P. M. and Lewis, D. (1975) A structural study of gallstones. Gut, 16, 630—637.

Bradbrook, R. A. and Bird, D. R. (1977) The clinical problems of small bowel enteroliths and diverticulosis. In preparation.

Bron, B., Waldram, R., Silk, D. B. A. and Williams, R. (1977) Serum, cerebrospinal fluid and brain levels of bile acids in patients with fulminant hepatic failure. Gut, 18, 692—696.

Campbell, C. B. and McIvor, W. E. (1975) A modified assay of total faecal bile acid excretion for clinical studies. Pathology, 7, 157—163.

Carey, Jr., J. B. (1958) The serum trihydroxy-dihydroxy bile acid ratio in liver and biliary tract disease. J. Clin. Invest., 37, 1494—1503.

Carey, Jr., J. B. (1973) Bile salt metabolism in man. In: The Bile Acids. Eds. P. P. Nair and D. Kritchevsky. Vol. 2, pp. 55—82. Plenum Press, New York—London.

Carey, M. C. and Small, D. M. (1974) Solubility of cholesterol in bile. In: Advances in Bile Acid Research. Eds. S. Matern, J. Hackenschmidt, P. Back and W. Gerok. pp. 277—283. F. K. Schattauer Verlag, Stuttgart—New York.

Cowen, A. E., Korman, M. G., Hofmann, A. F., Cass, O. W., Coffin, S. B. and Thomas, P. J. (1975a) Metabolism of lithocholate in healthy man. I. II. III. Gastroenterology, 69, 59—66, 67—76, 77—82.

Cowen, A. E., Korman, M. G., Hofmann, A. F. and Goldstein, N. P. (1975b) Biliary bile acid composition in Wilson's disease. Mayo Clinic Proc., 50, 229—233.

Dam, H., Kruse, I., Prange, I., Kallenhauge, H. E., Fenger, H. J. and Jensen, M. K. (1971) Studies on human bile. III. Composition of duodenal bile from healthy young volunteers compared with composition of bladder bile from surgical patients with and without uncomplicated gallstone disease. Z. Ernähr., 10, 160—177.

Danzinger, R. G., Hofmann, A. F., Schoenfield, L. J. and Thistle, J. L. (1972) Dissolution of cholesterol gallstones by chenodeoxycholic acid. N. Engl. J. Med., 286, 1—8.

Délèze, G., Karlaganis, G., Giger, W., Reinhard, M., Sidiropoulos, D. and Paumgartner, G. (1977) Identification of 3β-hydroxy-5-cholenoic acid in human amniotic fluid. In: Bile Acid Metabolism in Health and Disease. Eds. G. Paumgartner and A. Stiehl. pp. 59—62. MTP Press, Lancaster.

Demers, L. M. and Hepner, G. W. (1976) Levels of immunoreactive glycine-conjugated bile acids in health and hepatobiliary disease. Am. J. Clin. Pathol. 66, 831—839.

Dowling, R. H. (1976) Dissolution of cholesterol-gallstones I. In Falk Symposium No. 23, p. 34. Dr. Falk GmbH, Freiburg, i. Br., W. Germany.

Dyrszka, H., Salen, G., Zaki, F. G., Chen, T. and Mosbach, E. H. (1976) Hepatic toxicity in the Rhesus monkey treated with chenodeoxycholic acid for 6 months: biochemical and ultrastructural studies. Gastroenterology, 70, 93—104.

Eneroth, P. and Sjövall, J. (1971) Extraction, purification and chromatographic analysis of bile acids in biological materials. In: the Bile Acids. Eds. P. P. Nair and D. Kritchevsky. Vol. 1, pp. 121—171. Plenum Press, New York—London.

Englert, Jr., E., Harman, C. G. and Wales, Jr., E. E. (1969) Gallstones induced by normal foodstuffs in dogs. Nature (London), 224, 280—281.

Eyssen, H., Parmentier, G., Compernolle, F., Boon, J. and Eggermont, E. (1972) Trihydroxycoprostanic acid in the duodenal fluid of two children with intrahepatic bile duct anomalies. Biochim. Biophys. Acta, 273, 212—221.

Fausa, O. (1976) Serum bile acid concentrations after a test meal. Scand. J. Gastroent., 11, 229—232.

Fausa, O. and Gjone, E. (1976) Serum bile acid concentrations in patients with liver disease. Scand. J. Gastroent., 11, 537—543.

Fedorowski, T., Salen, G., Calallilo, A., Tint, G. S., Mosbach, E. H. and Hall, J. C. (1977) Metabolism of ursodeoxycholic acid in man. Gastroenterology, 73, 1131—1137.

Fischer, C. D., Cooper, N. S., Rothschild, M. A. and Mosbach, E. H. (1974) Effect of dietary chenodeoxycholic acid and lithocholic acid in the rabbit. Am. J. Dig. Dis., 19, 877—886.

Fisher, J. C., Bernstein, E. F. and Carey, Jr., J. B. (1965) Primary bile acid enterolith. Gastroenterology, 49, 272—279.

Freston, J. W. and Bouchier, I. A. D. (1968) Experimental cholelithiasis. Gut, 9, 2—4.

Fröhling, W. and Stiehl, A. (1976) Bile salt glucuronides: identification and quantitative analysis in the urine of patients with cholestasis. Eur. J. Clin. Invest. 6, 67—74.

Fromm, M. and Hofmann, A. F. (1971) Breath test for altered bile-acid metabolism. Lancet, 2, 621—625.

Gregg, J. A. (1967) Presence of bile acids in jaundiced human urine. Nature (London), 214, 29—31.

Grundy, S. M., Ahrens, E. H. Jr. and Miettinen, T. A. (1965) Quantitative isolation and gas-liquid chomatographic analysis of total fecal bile acids. J. Lipid Res., 6, 397—410.

Hanson, R. F., Isenberg, J. N., Williams, G. C., Hachey, D., Szczepanik, P., Klein, P. D. and Sharp, H. L. (1975) The metabolism of $3\alpha,7\alpha,12\alpha$-trihydroxy-5β-cholestan-26-oic acid in two siblings with cholestasis due to intrahepatic bile duct anomalies. J. Clin. Invest., 56, 577—587.

Haslewood, G. A. D. (1964) The biological significance of chemical differences in bile salts. Biol. Rev., 39, 537—574.

Haslewood, G. A. D. (1967) Bile Salts. Methuen, London.

Heaton, K. W. (1972) Bile Salts in Health and Disease. Churchill Livingstone, Edinburgh—London.

Heaton, K. W., Wicks, A. C. B. and Yeates, J. (1977) Bile composition in relation to race and diet. Studies in Rhodesian Africans and in British subjects. In: Bile Acid Metabolism in Health and Disease. Eds. G. Paumgartner and A. Stiehl. pp. 197—202. MTP Press, Lancaster.

Hegardt, F. G. and Dam, H. (1971) The solubility of cholesterol in aqueous solutions of bile salts and lecithin. Z. Ernähr, 10, 223—233.

Henegouwen, G. P., Van, B., Ruben, A. and Brandt, K. H. (1974) Quantitative analysis of bile acids in serum and bile, using gas-liquid chromatography. Clin. Chim. Acta, 54, 249—261.

Heywood, R., Palmer, A. K., Foll, C. V. and Lee, M. R. (1973) Pathological changes in fetal Rhesus monkey induced by oral chenodeoxycholic acid. Lancet, 2, 1021.

Hofmann, A. F., Simmonds, W. J., Korman, M. G., La Russo, N. F. and Hoffman, N. E. (1974) Radioimmunoassay of bile acids. In: Advances in Bile Acid Research. Eds. S. Matern, J. Hackenschmidt, P. Back and W. Gerok. pp. 95—98. F. K. Schattauer Verlag, Stuttgart—New York.

Holzbach, R. T., Marsh, M., Olszewski, M. and Holan, K. (1973) Cholesterol solubility in bile. Evidence that supersaturated bile is frequent in healthy man. J. Clin. Invest., 52, 1467—1479.

Holzbach, R. T., Corbusier, C., Marsh, M. and Naito, H. K. (1976) The process of cholesterol cholelithiasis induced by diet in the prairie dog; a physicochemical characterization. J. Lab. Clin. Med., 87, 987—998.

Howell, J. I., Lucy, J. A., Pirola, R. C. and Bouchier, I. A. D. (1970) Macromolecular assemblies of lipids in bile. Biochim. Biophys. Acta, 210, 1—6.

Igimi, H. (1976) Ursodeoxycholate — a common bile acid in gallbladder bile of Japanese subjects. Life Sci., 18, 993—1000.

Igimi, H., Tamesue, N., Ikejiri, Y. and Shimura, H. (1977) Ursodeoxycholate — *in vitro* cholesterol solubility and changes of human gallbladder bile after oral treatment. Life Sci., 21, 1373—1380.

Iser, J. H., Dowling, R. H., Mok, H. Y. I. and Bell, G. D. (1975) Chenodeoxycholic acid treatment of gallstones. N. Engl. J. Med., 293, 378—383.

Iser, J. H., Dowling, R. H., Murphy, G. M., de Leon, M., Ponz and Mitropoulos, K. A. (1977) Congenital bile salt deficiency associated with 28 years of intractable constipation. In: Bile Acid Metabolism in Health and Disease. Eds. G. Paumgartner and A. Stiehl. pp. 231—234. MTP Press, Lancaster.

Javitt, N. B. (1974) Bile salts and liver disease in childhood. Postgrad. Med. J., 50, 354—361.

Javitt, N. B., Lavy, U. and Kok, E. (1976) Bile acid balance studies in cholestasis. In: The Liver, Quantitative Aspects of Structure and Function. Eds. R. Preisig, J. Bircher and G. Paumgartner. pp. 249—254. Editio Cantor, Aulendorf, West Germany.

Kikuchi, H. (1972) The metabolic sequence for the occurrence of an anomalous bile acid, 12-ketochenodeoxycholic acid, found in the bile of hepatobiliary diseased patients. J. Biochem. (Tokyo), 72, 165—172.

Korman, M. G., La Russo, N. F., Hoffman, N. E. and Hofmann, A. F. (1975) Development of an intravenous bile acid tolerance test. N. Engl. J. Med., 292, 1205—1209.

Kritchevsky, D. and Story, J. A (1976) Dietary fiber and bile acid metabolism. In: The Bile Acids. Eds. P. P. Nair and D. Kritchevsky. Vol. 3, pp. 201—209. Plenum Press, London—New York.

La Russo, N. F., Hoffman, N. E., Hofmann, A. F. and Korman, M. G. (1975) Validity and sensitivity of an intravenous bile acid tolerance test in patients with liver disease. N. Engl. J. Med., 292, 1209—1214.

Little, J. M., Smallwood, R. A., Lester, R., Piasecki, G. J. and Jackson, B. T. (1975) Bile-salt metabolism in the primate fetus. Gastroenterology, 69, 1315—1320.

Low-Beer, T. S., Heaton, K. W., Pomare, E. W. and Reed, A. E. (1973) The effect of coeliac disease upon bile salts. Gut, 14, 204—208.

Makino, I., Hashimoto, H., Shinozaki, K., Yoshino, K. and Nakagawa, S. (1975) Sulfated and nonsulfated bile acids in urine, serum and bile of patients with hepatobiliary disease. Gastroenterology, 68, 545—553.

Makino, I. and Nakagawa, S. (1976) Changes of biliary bile acids in patients after administration of ursodeoxycholic acid. In: Bile Acid Metabolism in Health and Disease. Eds. G. Paumgartner and A. Stiehl. pp. 211—217. MTP Press, Lancaster.

Makino, I., Sjövall, J., Norman, A. and Strandvik, B. (1971) Excretion of 3β-hydroxy-5-cholenoic and 3α-hydroxy-5α-cholanoic acids in urine of infants with biliary atresia. Febs Letters, 15, 161—164.

Mann, W. N. (1974) Biliary cirrhosis: an appraisal. Guy's Hosp. Reports, 123, 197—219.

Marks, J. W., Bonorris, G. G. and Schoenfield, L. J. (1976) Pathophysiology and dissolution of cholesterol gallstones. In: The Bile Acids. Eds. P. P. Nair and D. Kritchevsky. Vol. 3, pp. 81—113. Plenum Press, London—New York.

Mashige, F., Imai, K. and Osuga, T. (1976) A simple and sensitive assay of total serum bile acids. Clin. Chim. Acta, 76, 79—86.

Matern, S., Haag, M., Hans, Ch. and Gerok, W. (1977) Oral cholate tolerance test — an application of specific radioimmunoassays for the determination of serum-conjugated cholic and deoxycholic acids. In: Bile Acid Metabolism in Health and Disease. Eds. G. Paumgartner and A. Stiehl. pp. 253—261. MTP Press, Lancaster.

Matern, S., Krieger, R. and Gerok, W. (1976) Radioimmunoassay of serum conjugated cholic acid. Clin. Chim. Acta, 72, 39—48.

McSherry, C. K., Javitt, N. B., de Carvalho, J. M. and Glenn, F. (1971) Cholesterol gallstones and the chemical composition of bile in baboons. Ann. Surg., 173, 569—577.

Maton, P. N., Murphy, G. M. and Dowling, R. H. (1977) Ursodeoxycholic acid treatment of gallstones: dose response study and possible mechanism of action. Lancet, 2, 1297—1301.

Mok, H. Y. I., Bell, G. D. and Dowling, R. H. (1974) Effect of different doses of chenodeoxycholic acid on bile-lipid composition and on frequency of side-effects in patients with gallstones. Lancet, 2, 253—257.

Mok, H. Y. I. and Dowling, R. H. (1974) How well can we measure bile acid pool size? In: Advances in Bile Acid Research. Eds. S. Matern, J. Hackenschmidt, P. Back and W. Gerok. pp. 315—323. F. K. Schattauer Verlag, Stuttgart—New York.

Morrisey, K. P., McSherry, C. K., Swarm, R. L., Nieman, W. H. and Dietrick, J. E. (1975) Toxicity of chenodeoxycholic acid in the nonhuman primate. Surgery, 77, 851—860.

Murphy, G. M. (1972) Serum bile acids in cholestatic liver disease: their measurement and significance. Ann. Clin. Biochem., 9, 67—73.

Murphy, G. M., Edkins, S. M., Williams, J. W. and Catty, D. (1974) The preparation and properties of an antiserum for the radioimmunoassay of serum conjugated cholic acid. Clin. Chim. Acta, 54, 81—89.

Nakagawa, S., Makino, I., Ishizaki, T. and Dohi, I. (1977) Dissolution of cholesterol gallstones by ursodeoxycholic acid. Lancet, 2, 367—369.

Nakayama, F. and van der Linden, W. (1970) Bile from gallbladder harbouring gallstone: can it indicate stone formation? Acta. Chir. Scand., 136, 605—610.

Northfield, T. C. (1973) Intraluminal precipitation of bile acids in stagnant loop syndrome. Br. Med. J., 743—745.

Northfield, T. C., Drasar, B. S. and Wright, J. T. (1973) Value of small intestinal bile acid analysis in the diagnosis of the stagnant loop syndrome. Gut, 14, 341—347.

Northfield, T. C., La Russo, N. F., Hofmann, A. F. and Thistle, J. L. (1975) Effect of chenodeoxycholic acid treatment in gallstone subjects. Gut, 16, 12—17.

Northfield, T. C. and McColl, I. (1973) Postprandial concentrations of free and conjugated bile acids down the length of the normal human intestine. Gut, 14, 513—518.

Oviedo, M. A., Ho, K.-J., Bliss, K., Soong, S.-J., Mikkelson, B. and Taylor, C. B. (1977) Gallbladder bile composition in different ethnic groups. Arch. Pathol. Lab. Med., 101, 208—212.

Palmer, R. H. (1972) Bile acids, liver injury and liver disease. Arch. Intern. Med., 130, 606—617.

Pedersen, L. and Bremmelgaard, A. (1976) Hepatic morphology and bile acid composition of bile and urine during chenodeoxycholic acid therapy for radiolucent gallstones. Scand. J. Gastroent., 11, 385—389.

Pomare, E. W., Heaton, K. W., Low-Beer, T. S. and Espiner, H. J. (1976) The effect of wheat bran upon bile salt metabolism and upon the lipid composition of bile in gallstone patients. Digestive Diseases, 21, 521—526.

Salen, G. and Mosbach, E. H. (1976) The metabolism of sterols and bile acids in cerebrotendinous xanthomatosis. In: The Bile Acids. Eds. P. P. Nair and D. Kritchevsky. Vol. 3, pp. 115—153. Plenum Press, London—New York.

Scarpello, J. H. B. and Sladen, G. E. (1977) ^{14}C-glycocholate test in Crohn's disease etc., Gut, 18, 736—748.

Schaffner, F., Bacchin, P. G., Hutterer, F., Scharnbeck, H. H., Sarkozi, L. L., Denk, H.

and Popper, H. (1971) Mechanism of cholestasis. 4. Gastroenterology, 60, 888—897.
Scherstén, T. (1967) The synthesis of cholic acid conjugates in human liver. Acta Chir. Scand. Supplementum 373.
Schoenfield, L. J. and Sjövall, J. (1966) Bile acids and cholesterol in guinea pigs with induced gallstones. Am. J. Physiol., 211, 1069—1074.
Schoenfield, L. J., Coyne, M. J., Bonorris, G. G., Key, P. H. and Marks, J. W. (1976) Dissolution of cholesterol-gallstones II. In Falk Symposium No. 23, p. 35. Dr. Falk GmbH, Freiburg i. Br., West Germany.
Schwarz, H. P., Bergmann, K. V. and Paumgartner, G. (1974) A simple method for the estimation of bile acids in serum. Clin. Chim. Acta, 50, 197—206.
Schwenk, M., Hofmann, A. F., Carlson, G. L., Carter, J. A., Coulston F. and Greim, H. (1978) Bile acid conjugation in chimpanzee: effective sulfation of lithocholic acid. Arch. Toxicol., 40, 109—118.
Sheltawy, M. J. and Losowsky, M. S. (1975) Determination of faecal bile acids by an enzymic method. Clin. Chim. Acta, 64, 127—132.
Stiehl, A. (1974) Bile salts sulphates in cholestasis. Europ. J. Clin. Invest., 4, 59—63.
Stiehl, A., Earnest, D. L. and Admirand, W. H. (1975) Sulfation and renal excretion of bile salts in patients with cirrhosis of the liver. Gastroenterology, 68, 534—544.
Subbiah, M. T. R., Marai, L., Dinh, D. M. and Penner, J. W. (1977) Sterol and bile acid metabolism during development. 1. Studies on the gallbladder and intestinal bile acids of newborn and fetal rabbit. Steroids, 29, 83—92.
Summerfield, J. A., Billing, B. H. and Shackleton, C. H. L. (1976a) Identification of bile acids in the serum and urine in cholestasis. Biochem. J., 154, 507—516.
Summerfield, J. A., Gollan, J. L. and Billing, B. H. (1976b) Synthesis of bile acid monosulphates by the isolated perfused rat kidney. Biochem. J., 156, 339—345.
Summerfield, J. A., Cullen, J., Barnes, S. and Billing, B. H. (1977) Evidence for renal control of urinary excretion of bile acids and bile acid sulphates in the cholestatic syndrome. Clin. Sci. Mol. Med., 52, 51—65.
Sutor, D. J. and Wooley, S. E. (1971; 1973; 1974) Gut, 12, 55—64; 14, 215—220; 15, 130—131 and 487—491.
Sutor, D. J., Wooley, S. E. and Gordon, E. M. (1969) Composition of gallstones from hamsters. Gut, 10, 684.
Thaysen, E. H. and Pedersen, L. (1977) Clinical evaluation of the [^{14}C] cholylglycine breath test. In: Bile Acid Metabolism in Health and Disease. Eds. G. Paumgartner and H. Stiehl. pp. 263—269. MTP Press, Lancaster.
Thistle, J. L. and Schoenfield, L. J. (1969) Induced alteration of bile composition in humans with cholelithiasis. J. Lab. Clin. Med., 74, 1020—1021.
Thjodleifsson, B., Barnes, S., Chitranukroh, A., Billing, B. H. and Sherlock, S. (1977) Assessment of the plasma disappearance of cholyl-1 ^{14}C-glycine as a test of hepatocellular disease. Gut, 18, 697—702.
Thomas, P. J. and Hofmann, A. F. (1973) A simple calculation of the lithogenic index of bile: expressing biliary lipid composition on rectangular coordinates. Gastroenterology, 65, 698—700.
Toouli, J., Jablonski, P. and Watts, J. McK. (1975) Gallstone dissolution in man using cholic acid and lecithin. Lancet, 2, 1124—1126.
Vlahcevic, Z. R., Bell, C. E. Jr., Buhac, I., Farrar, J. T. and Swell, L. (1970) Diminished bile acid pool size in patients with gallstones. Gastroenterology, 59, 165—173.

Van der Linden, W. and Norman, A. (1967) Composition of human hepatic bile. Acta Chir. Scand., 133, 307—313.

Weber, A. M., Roy, C. C., Morin, C. L. and Lasalle, R. (1973) Malabsorption of bile acids in children with cystic fibrosis. N. Engl. J. Med., 289, 1001—1005.

Webster, K. H., Lancaster, M. C., Hofmann, A. F., Wease, D. F. and Baggentoss, A. H. (1975) Influence of primary bile acid feeding on cholesterol metabolism and hepatic function in the Rhesus monkey. Mayo Clinic. Proc., 50, 134—138.

Wood, J. R., France, V. M. and Sutor, D. J. (1974) Occurrence of gallstones in foetal sheep. Lab. Animals, 8, 155—159.

Yoshida, T., McCormick, W. C., Swell, L. and Vlahcevic, Z. R. (1975) Bile acid metabolism in cirrhosis. Gastroenterology, 68, 335—341.

CHAPTER 6

General conclusions and speculations

The purpose of the previous chapters was to put the reader into the biological picture (physiological, biochemical, taxonomic and medical) as far as it concerns bile salts. I would like to conclude with some largely speculative remarks about the significance of this picture, particularly in relation to vertebrate evolution.

It seems possible that there is a sharp division between vertebrate and invertebrate animals at the level of hagfishes, for no animal with less organisation as a chordate than these has yet been shown to make use of steroid detergent digestive molecules. However, some of those non-arthropod invertebrate forms that can synthesize cholesterol (Chapter 4) might have digestive fluids containing steroid bile salts or biochemical precursors of these substances. A sea pen, *Ptilosarcus gurneyi*, of the order Pennatulacea, contains methyl esters of steroid acids with the cholesterol nucleus and C_{24} bile acid side-chains, but with S instead of R steric relationships at C-20 (e.g. Vanderah and Djerassi, 1977). Such discoveries show that some invertebrates can shorten the cholesterol side-chain, but do not of themselves point to these animals as likely remnants of stock that gave rise to chordates, for the newly-found steroids are not detergent-type molecules. Hagfishes themselves are the specialised remnants of vertebrates whose ancestors very long ago departed from invertebrates. At present, therefore, the steroid bile salts cannot tell us which existing invertebrates, if any, might be the present-day descendents of the stock that gave rise to vertebrates. This question has been raised before in comparative biochemistry particularly in relation to the phosphagens of muscle tissue, and with the same inconclusive answer.

Perhaps the most astonishing thing about vertebrate bile salts is their variety. Each of the chemical types described in Chapter 2 seems to be able to function perfectly well physiologically as a digestive aid in appropriate animals and the difficulty is to understand why evolutionary development took place at all. The only known bile salt molecules not obviously well-structured to act as effective amphiphiles are the disulphates of myxinol and 16-deoxymyxinol, yet hagfishes do very well with these substances (albeit in large amounts), as they do with other molecules and morphological features that are apparently less effective than those of more advanced forms. What selective pressures, then, led to an advance from such poor amphiphiles as hagfish bile salts to the clearly superior (as detergents) side-chain alcohol monosulphates having the $3\alpha,7\alpha,12\alpha$-trihydroxy nucleus with additional OH groups, from these to taurine-conjugated C_{27} acids, thence to the taurine conjugates of C_{24} acids and finally to the glycine conjugates of eutherian mammals?

Perhaps we can set aside hagfishes as remarkable examples of obsolete models surviving in a modernised animal world and *Latimeria*, too, with its 3β-hydroxyl group in latimerol, might come into an analogous relict category. This still leaves the great variety in the remaining vertebrates to be accounted for, for we can by no means say that the polyhydroxy alcohol sulphates are inferior as amphiphiles to taurine conjugates of C_{27} or C_{24} bile acids. Glycine conjugation is even more difficult to explain, for if it confers selective advantages why do we not see it in other highly advanced vertebrates: in teleosts, lizards, snakes, birds or marsupials?

And what about evolutionary "drift"? It seems impossible to believe that cholic acid has not been evolved separately a number of times: in bony fishes, amphibians, reptiles, birds and mammals. Even in elasmobranchs, there are signs of its appearance. The alternative that teleosts, most lizards and snakes, birds and mammals arose from precursors already having cholic acid seems quite untenable, from the bile salt distribution given in Chapter 4. Why should evolution proceed separately to the same substance and its C_{24} allies in every vertebrate group?

Is it possible that an answer lies in the nature of evolution itself? That once a process of biochemical evolution has started there is a tendency, unless it is prevented from so doing by selective forces, for it to proceed to an end finally determined by the functional molecule and its place in the entire phenotypic biochemical scheme? To approach this question experimentally in our case, we need to know the mechanisms by which expression of the genes coding for enzymes making bile salts from cholesterol are controlled. Are the DNA sequences coding for all the enzymes acting in advanced animals already present in the most primitive vertebrates and does evolution merely entail that they should be expressed as protein and this expression controlled? Perhaps the situation is analogous to that in the specialised tissues of a single animal, which we know to have the genetic potential to make any metabolite made anywhere in the organism but which do not actually do so. When more is known about how DNA expression is determined in different tissues and different animal forms, perhaps a fresh look at bile salts will give a new insight into evolutionary progress.

There is another general question, perhaps no more than a rider to the one discussed above and approachable experimentally in a similar way, and that is: what is the biochemical meaning of evolutionary vigour and senescence? How is it that some animal groups, for example crocodiles, have changed little over very long periods whereas others, for instance ranid frogs, are apparently in a state of active evolutionary radiation? Is there such a thing as an innate capacity to evolve in response to changing circumstances, that can be lost perhaps for ever? The bile salts of crocodiles and some frogs fit very well with the evolutionary status, senescent or vigorous, of the animals as assessed above and a more intimate study of the mechanisms producing them might eventually throw light on this question of evolutionary potential.

A striking feature of steroid bile salts is the obviously close relationship between their distribution amongst vertebrates and the complete biosynthetic pathway leading to C_{24} acids, as set out in Scheme 3.1. If this relationship is shown as in Scheme 6.1, it can account for almost all known major bile alcohols or bile acids as arising by one or two simple biochemical steps from substances on the complete chief biochemical pathways. Myxinol and latimerol are exceptions, for we do not know how the 3β OH of cholesterol is retained or restored.

Apart from these two primitive bile alcohols, $7\alpha,12\alpha$-dihydroxy-4-cholesten-3-one (3.3) is a common intermediate, lying at the junction between the 5α and 5β series. We do not know how the cholesterol side-chain is shortened to form petromyzonol, but otherwise the relationships between biosynthetic intermediates and functioning

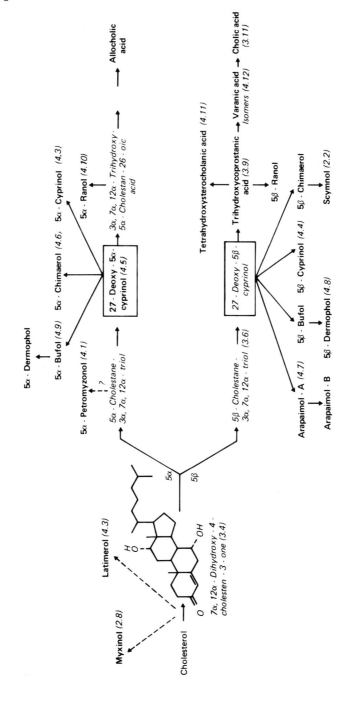

Scheme 6.1. Probable relationships between the chief biosynthetic pathways for C_{24} bile acids and the bile acids and alcohols described in Chapter 4. Known principal bile acids and alcohols shown in bold type. Chenodeoxycholic acid lies on the same pathway as cholic acid and the other chief primary C_{24} acids are derived from chenodeoxycholic or cholic acids.

bile salts is biochemically very close and seems to me to emphasize the point made above, that what we are studying may be the gradual expression of DNA sequences already present in primitive vertebrates.

Recent discoveries of large amounts of trihydroxycoprostanic acid (3.9) in some human infants with biliary atresia and of C-25 hydroxylated alcohols resembling 5β-bufol in human cerebrotendinous xanthomatosis (pp. 53 and 140) show again that man still retains the biochemical capacity to make more primitive types of bile salts, a situation that has implications not only for human genetics but also for evolutionary studies.

New methods that can discover and identify small traces of compounds in the complex mixtures of bile alcohols or bile acids obtained after solvolysis or hydrolysis of native bile salts have made it very clear that most animal species normally make a considerable variety of bile salt molecules, of which only a few could be physiologically important. In addition to these chief bile salts, the minor constituents, described as "biochemical noise", perhaps indicate some of the possibilities of biological variation. The "noise" may seem senseless, as in the mixtures of relict bile alcohols still to be found in sturgeons or polypterid fishes that have mainly C_{24} taurine-conjugated acids, or it may appear to herald incipient modernization like the minor amounts of cholic and allocholic acids in amphibians and cyprinid fishes. In human disease such normally minor bile salts may become prominent, as mentioned above. Their study is, I think, important as telling us what genes in each case can still be expressed: their interpretation as relict or potentially useful in the future is based on a view of the complete evolutionary picture and of course may be mistaken.

Apart from our work, the philosophy behind which has been set out in this and earlier Chapters, there are other published attempts to make biological sense of the differences between vertebrate bile salts. T. Hoshita and his colleagues at Hiroshima have pursued both the detailed chemistry and species distribution particularly in amphibians with a declared intention of understanding "the biosynthesis and molecular evolution of bile acid" (Kihira et al., 1977).

Ripatti and Sidorov (1973) attempted a correlation between the diet and the bile acids of 13 species of fish, 13 birds and 7 eutherian mammals; they concluded that their findings supported "to a substantial degree the existence of a relationship between the type of nutrition of the animal and the qualitative and quantitative composition of the bile acids in the bile".

Yousef et al. (1973) compared the bile acid composition of 7 species of desert rodents and also tried to relate this to diet. They thought "sterol metabolism differs in various species and this difference does not appear to be related to ecology, nutrition or ecological distribution". In later work the same authors (Yousef et al., 1977) studied by quantitative GLC—MS the bile acid composition in 9 lizard species. They describe a "similarity index" based on this composition and used it to discuss taxonomic relationships. As Yousef et al. clearly realise, the biological value of such comparisons must be assessed with great caution, since so many very different animals share a few molecules as bile salts, some of these molecules are secondary and there has been a great deal of convergent and parallel evolution. Again, the Yousefs and their colleagues could find no correlation between diet and bile salt chemistry.

It seems to me to be unlikely, if not impossible, that a single biochemical character could be used to construct a valid scheme of evolutionary relationships between animal species: the assumption by biochemists that such a thing could be done seems naive and likely to bring comparative biochemistry into contempt. It is the business of systematists to propose such schemes and in so doing they must be expected to take molecular characters into account. Molecular differences are precise and can be directly related to genes; for these reasons alone they should be highly rated by systematists. Each biochemical character should be considered in its context as contributing to the animals' physiology and with as much understanding as possible of the selective pressures that have influenced its history.

I am aware that some of the ideas put forward here are not new; indeed they constitute, with different degrees of credibility, part of the general thinking concerning evolution. It is precisely because these thoughts, arising as they do from a molecular study, coincide so well with opinions based mainly on morphology and palaeontology that I have come to believe that bile salt variation may be an excellent character for deeper investigation. For molecules make up the machinery of living processes. A complete molecular understanding of these processes is possible and might tell us how they came to have the results we see in living forms today and might expect in the future. Odd as it seems, one possible route to such a goal may be through a closer study of bile salts.

References

Kihara, K., Yasuhara, M., Kuramoto, T. and Hoshita, T. (1977) New bile alcohols, 5α- and 5β-dermophols from amphibians. Tetrahedron Lett., 8, 687—690.

Ripatti, P. O. and Sidorov, V. S. (1973) Quantitative composition of the bile acids of certain vertebrates in connection with the nature of their nutrition. Dokl. Akad. Nauk. SSSR., 212, 770—773.

Vanderah, D. J. and Djerassi, C. (1977) Novel marine sterols with modified bile acid side chain from the sea pen *Ptilosarcus gurneyi*. Tetrahedron Lett., 8, 683—686.

Yousef, I. M., Yousef, M. K. and Bradley, W. G. (1973) Bile acid composition in some desert rodents. Proc. Soc. Expl. Biol. Med., 143, 596—601.

Yousef, I. M., Bradley, W. G. and Yousef, M. K. (1977) Bile acid composition of some lizards from southwestern United States. Proc. Soc. Expl. Biol. Med., 154, 22—26.

APPENDIX

Bile salts in different animal forms

Throughout our studies, my colleagues and I have tried to publish records of all bile salts discovered. The Table below is intended to list those not given by Haslewood (1967) or Tammar (1971a, b, c). Classification of fishes is mainly that of Greenwood (1975).

Except for some recent work, only the chief bile acids or alcohols have been identified for gas-chromatographic-mass spectral methods are quite new and have not generally been used for bile salt analysis.

APPENDIX

Higher animal group	Species	Bile alcohols; bile acids*	Type of conjugate	References
PROTOCHORDATES	*Branchiostoma* (*Amphioxus*) *lanceolatum*	Probably not steroids		Chapter 4 (p. 80)
VERTEBRATES Fishes				
CHONDRICHTHYES	*Squalus acanthias* and *Raia clavata*	Scymnol; 5β-chimaerol; no more than traces of any other bile alcohol	Sulphate	U.o. (mild solvolysis and GLC)
TELEOSTOMI CROSSOPTERYGII	*Latimeria chalumnae*	**Latimerol**; 5α-bufol; 5α-cyprinol; unidentified alcohols; traces of acids	Sulphate	
DIPNOI	*Lepidosiren paradoxa*	5α-**Bufol**; 5α-cyprinol; **27-deoxy-5α-cyprinol**; unidentified alcohols; traces of acids	Sulphate	Amos et al., 1977
	Neoceratodus forsteri	5α-Bufol; 5α-**chimaerol**; 5α-petromyzonol, 26-deoxy-5α-chimaerol; **27-deoxy-5α-cyprinol** unidentified alcohols; traces of acids	Sulphate	
	Protopterus aethiopicus	5α-**Cyprinol**; 5α- and 5β-petromyzonol; 27-deoxy-5α-cyprinol; unidentified alcohols; traces of acids	Sulphate	
BRACHIOPTERYGII	*Polypterus congicus*	Cholic acid; chenodeoxycholic acid; deoxycholic acids; 5β-cyprinol	Taurine; sulphate	
	Polypterus ornatipinnis	Cholic acid; deoxycholic acid; unidentified C_{24} acid; 5β-cyprinol	Taurine; sulphate	U.o.[1]
	Polypterus senegalus	Cholic acid; haemulcholic acid; deoxycholic acid; unidentified C_{24} acid; 5α-chimaerol	Taurine; sulphate	
	Calomoichthys calabricus	Generally similar to *Polypterus*		U.o.

APPENDIX (continued)

Higher animal group	Species	Bile alcohols; bile acids*	Type of conjugate	References
TELEOSTEI				
OSTEOGLOSSIFORMES	*Mormyrus caballus*	Haemulcholic acid; cholic acid; unidentified C_{24} acids	Taurine	U.o.[1]
	Scleropages leichardtii	Cholic acid; other acids	Taurine	U.o.
SALMONIFORMES				
Salmonidae	*Salmo salar, S. trutta, Coregonus albula, C. lavaretus, C. muksum, C. nasus, C. peled, Stenodus leucichthys nelma Thymallus thymallus*	Cholic acid; chenodeoxycholic acid		Ripatti and Sidorov, 1973
	Rainbow trout, *Salmo gairdneri*	Cholic (85%), chenodeoxycholic (about 14%), $3\alpha,12\alpha$-dihydroxy-7-oxo- and $7\alpha,12\alpha$-dihydroxy-3-oxo-5β-cholanic acids	Taurine? glycine; sulphate	Denton et al., 1974
OSTARIOPHYSI				
Characoidei	*Alestes imberi*	Chenodeoxycholic and other acids; traces of alcohols	Taurine; sulphate	
	Alestes macrolepidotus			
	Alestes macrophthalmus			
	Distichodus lususso	Cholic, chenodeoxycholic and other acids; traces of alcohols	Taurine; sulphate	U.o.[1]
	Distichodus maculatus			
	Distichodus sexfasciatus			
	Tigerfish, *Hydrocynus vittatus*	Cholic, chenodeoxycholic and other acids	Taurine	
Gymnotoidei	*Electrophorus electricus*	Cholic acid; prob. deoxycholic acid; other acids	Taurine	U.o.
Cyprinoidei				
Cyprinidae	*Barbus eutaenia, B. trachypterus* and *B. gestetneri.*	5α-Cyprinol and other alcohols; minor amounts of acids	Sulphate; taurine	U.o.[1]
	Varichorhinus upembensis	5α-Cyprinol; 27-deoxy-5α-cyprinol; other alcohols; minor amounts of acids	Sulphate; taurine	
	Labeo cf. *chariensis* and *L. velifer*	5α-Chimaerol; other alcohols	Sulphate	U.o.[1]

APPENDIX (continued)

Higher animal group	Species	Bile alcohols; bile acids*	Type of conjugate	References
Catostomidae	Carpsucker, *Carpoides carpio*	Allocholic acid; ? 5α-cyprinol	Taurine; Sulphate	Briggs and Bussjaeger, 1972
	Suckers, *Catostomus macrocheilus* and *C. plebius*, *Chasmistes brevirostris*	5α-Chimaerol	Sulphate	
	Blue sucker, *Cycleptus elongatus*	5α-Chimaerol (65%); 5α-cyprinol (35%)	Sulphate	
	Hogsucker, *Hypentelium nigricans*	5α-Chimaerol (74%); 5α-cyprinol (22%)	Sulphate	
	Buffalo fish, *Ictobius bubalus* and *I. cyprinellus*	5α-Cyprinol	Sulphate	Bussjaeger, 1973
	Sucker, *Minytrema melanops*	5α-Chimaerol (49%); 5α-cyprinol (51%)	Sulphate	
	Redhorses, *Moxostoma carinatum*, *M. duguesnei* and *M. erythrurum*	5α-Chimaerol (80–87%); 5α-cyprinol (13–20%)	Sulphate	
SILURIFORMES	*Ictalurus furcatus* and *I. punctatus*	**Cholic acid**; chenodeoxycholic acid; deoxycholic acid (*I. punctatus*)	Taurine	Kellogg, 1975
	Schilbe mystus	Cholic, haemulcholic, chenodeoxycholic, deoxycholic and other acids; prob. 5β-cyprinol (minor amount)	Taurine; sulphate	
	Heterobranchus longifilis	Cholic, chenodeoxycholic and other acids	Taurine	
	Malapterurus electricus	Haemulcholic acid, cholic acid; other acids	Taurine	U.o.[1]
	Chrysichthys sp.	Cholic acid, deoxycholic acid, haemulcholic and other acids	Taurine	
	Synodontis sps.	Unconjugated C_{24} acids; taurine conjugates		
GADIFORMES	Burbot, *Lota lota*	Cholic acid (about 90%)		Ripatti and Sidorov, 1973
PERCIFORMES	Perch, *Perca fluviatilis*	Cholic acid (about 85%)		

APPENDIX (continued)

Higher animal group	Species	Bile alcohols; bile acids*	Type of conjugate	References
Amphibians				
APODA	Caecilian, *Dermophis mexicanus*	5β-Dermophol	Sulphate	Kihara et al., 1977
CAUDATA				
Amphiumidae	*Amphiuma means*	5α-Bufol (93%); 5α-dermophol (6%)	Sulphate	
Salamandridae	*Diemyctylus pyrrhogaster*	5α-Bufol (60%); 5α-cyprinol (8%); 5α-dermophol (28%)	Sulphate	
	Megalobatrachus japonicus	5α-Cyprinol (64%); 5α-dermophol (16%)	Sulphate	
	Necturus maculosus	5α-Cyprinol; 3α,7α,12α-trihydroxy-5α-cholestan-26-oic acid; allocholic acid; cholic acid	Sulphate; taurine	Ali, 1975; u.o.
ANURA				
Discoglossidae	*Discoglossus pictus*	5β-Bufol; 5β-cyprinol; 5β-ranol; **26-deoxy-5β-ranol; trihydroxycoprostanic acid**; cholic acid	Sulphate, taurine (some unconjugated acids)	Anderson et al., 1974
	Bombina orientalis	Trihydroxycoprostanic acid and an isomer of varanic acid	None	Kuramoto et al., 1973
Pipidae	*Xenopus laevis*	**5β-Cyprinol; trihydroxycoprostanic**, cholic and chenodeoxycholic acids	Sulphate	Kuramoto et al., 1973; Anderson et al., 1974
	Xenopus mulleri	5β-Cyprinol; 27-deoxy-5β-cyprinol; 26-deoxy-5β-ranol; trihydroxycoprostanic acid	Sulphate	Anderson et al., 1974
Bufanidae	*Bufo b. formosus*	5β-Bufol (90%); 27-deoxy-5α- and -5β-cyprinol (2% each); 5β-dermophol (3%); trihydroxybufosterocholenic acid; other steroids	Sulphate	Anderson et al., 1974; Kihara et al., 1977
	Bufo b. vulgaris	5β-**Bufol**; 27-deoxy-5α-cyprinol; 27-deoxy-5β-cyprinol; trihydroxycoprostanic acid (unconjugated) other steroids	Sulphate	Anderson et al., 1974

APPENDIX (continued)

Higher animal group	Species	Bile alcohols; bile acids*	Type of conjugate	References
	Bufo maculatus, B. marinus; B. regularis	5β-**Bufol**; 27-deoxy-5β-cyprinol; unconjugated trihydroxycoprostanic acid	Sulphate	Anderson et al., 1974
Hylidae	Hyla arborea japonica	5β-Ranol	Sulphate	Kuramoto et al., 1973
Ranidae	Arthroleptis ? poecilonotus	5β-Bufol; 5β-cyprinol; 5β-ranol; 27-deoxy-5α- and -5β-cyprinol; 26-deoxy-5β-ranol; 26-deoxy-26-nor-5β-ranol; trihydroxycoprostanic acid	Sulphate; taurine	Anderson et al., 1974
	Cardioglossa leucomystax	27-Deoxy-5β-cyprinol; 26-deoxy-5α- and -5β-ranol; arapaimol-A; unidentified steroids	Sulphate	
	Hylarana albolabris	5α-Bufol; 5α-cyprinol; 5β-ranol; 26-deoxy-5α-ranol	Sulphate	
	Aubria subsigillata	5β-Ranol; 26-deoxy-5β-ranol; 26-deoxy-26-nor-5β-ranol; unconjugated trihydroxycoprostanic acid	Sulphate	
	Dicroglossus occipitalis	5β-Bufol; 5β-cyprinol; 27-deoxy-5α- and -5β-cyprinol; 26-deoxy-5α- and -5β-ranol; 26-deoxy-26-nor-5β-ranol; trihydroxycoprostanic acid (free and conjugated); unidentified steroids	Sulphate; taurine	
	Ptychadena acquiplicata	5β-Bufol; 5β-cyprinol; 5β-ranol; 27-deoxy-5α- and -5β-cyprinol; 26-deoxy-5α- and -5β-ranol	Sulphate	Anderson et al., 1974
	Ptychadena longirostris and P. maccarthyensis	5β-Bufol; 5β-cyprinol; 27-deoxy-5β-cyprinol; 26-deoxy-5α- and -5β-ranol; unidentified steroids	Sulphate	

APPENDIX (continued)

Higher animal group	Species	Bile alcohols, bile acids*	Type of conjugate	References
	P. maccarthyensis tadpoles	Different from the adult pattern	Sulphate	U.o.[2]
	Ptychadena mascareniensis	5α- and 5β-Bufol; 5α- and 5β-cyprinol; 27-deoxy-5α- and -5β-cyprinol; **26-deoxy-5α- and -5β-ranol**	Sulphate	Anderson et al., 1974
	P. mascareniensis tadpoles	Different from the adult pattern	Sulphate	U.o.[2]
	Ptychadena oxyrhynchus	5β-Bufol; **5β-cyprinol**; 27-deoxy-5β-cyprinol; 26-deoxy-5β-ranol; unconjugated trihydroxycoprostanic acid	Sulphate	Anderson et al., 1974
	P. oxyrhynchus tadpoles	Mainly unconjugated trihydroxycoprostanic acid		U.o.[2]
	Ptychadena taenioscelis	5α-**Bufol**; 5α-cyprinol; 27-deoxy-5α-cyprinol; **26-deoxy-5α-ranol**; unidentified steroids	Sulphate	Anderson et al., 1974
	Rana brevipoda	5β-Cyprinol; trihydroxycoprostanic and cholic acids (free and conjugated)	Sulphate; taurine	Kuramoto et al., 1973
	Rana catesbeiana	5α-Cyprinol; 5α- and 5β-ranol; **26-deoxy-5α- and -5β-ranols**, isomeric at C-24	Sulphate	Noma et al., 1976
	Rana catesbeiana tadpoles (up to Stage 18)	Much more 26-deoxy-5β-ranol than later stages	Sulphate	U.o.[3]
	Rana esculenta	5α- and 5β-Cyprinol; unidentified steroids	Sulphate	Anderson et al., 1974
	Rana galamensis	5α-Bufol; **5α-cyprinol**; 26-deoxy-5α-ranol	Sulphate	Anderson et al., 1974
	Rana limnocharis limnocharis	5β-**Bufol**; trihydroxycoprostanic and cholic acids (free and conjugated)	Sulphate; taurine	Kuramoto et al., 1973

APPENDIX (continued)

Higher animal group	Species	Bile alcohols; bile acids*	Type of conjugate	References
	Rana pipiens	5β-Ranol; 26-deoxy-5β-ranol; 26-deoxy-26- nor 5α- and 5β-ranol; unidentified steroids	Sulphate	Anderson et al., 1974
	Rana ridibunda	5β-Cyprinol; 27-deoxy-5β-cyprinol; trihydroxycoprostanic acid	Sulphate; taurine	Kuramoto et al., 1973
	Rana rugosa	5α-Cyprinol; trihydroxycoprostanic acid	Sulphate; taurine	Anderson et al., 1974
	Rana temporaria	5α-Ranol; 26-deoxy-5α-ranol	Sulphate	Ali, 1975
	Rana tigrina	Trihydroxycoprostanic, allcoholic, cholic and deoxycholic acids		
Rhacophoridae	Phylctimantis leonardi	Tadpoles different from adults		U.o.²
	Hyperolius concolor ibadensis and H. fusciventris burtoni	5β-Bufol; 5β-cyprinol; 27-deoxy-5β-cyprinol; arapaimol-A (H.f.b. only); unidentified steroids	Sulphate	Anderson et al., 1974
	Hyperolius guttulatus	5β-Bufol; 5β-cyprinol; 27-deoxy-5α- and -5β-cyprinol; arapaimol-A; unidentified steroids	Sulphate	
	Hyperolius nitidulus	5β-Cyprinol; 27-deoxy-5β-cyprinol; arapaimol-A; unidentified steroids	Sulphate	
Reptiles				
TESTUDINES	Green turtle, Chelonia mydas	**Tetrahydroxysterocholanic acid;** 3α,12α,22ξ-trihydroxy-5β-cholestane-26(or 27)-oic acid; cholic and deoxycholic acids	Taurine	Haslewood et al., 1978

APPENDIX (continued)

Higher animal group	Species	Bile alcohols; bile acids*	Type of conjugate	References
SAURIA				
Gekkonidae	*Coleonyx variegatus*	**Cholic**, **allocholic**, chenodeoxycholic, deoxycholic and lithocholic acids	Taurine	Yousef et al., 1977
Iguanidae	*Dipsosaurus dorsalis*	Cholic, **allocholic**, deoxycholic and lithocholic acids	Taurine	
	Phrynosoma platyrhinos; Scleroporus graciosus; S. magister; S. occidentalis; Urosaurus graciosus; Uta stansburiana	Cholic, **allocholic**, chenodeoxycholic, deoxycholic and lithocholic acids	Taurine	
	Iguana iguana	**Allocholic** and $3\alpha,7\alpha,12\alpha$-trihydroxy-5α-cholestan-26-oic acids		Okuda et al., 1972
Agamidae	*Uromastix hardwickii; U. microlepis*	Allocholic acid (about 90%); deoxycholic acid (U.h.); other 5α- and 5β-acids	Taurine	Ali et al., 1976; U.o.
Teiidae	*Cnemidophorus tigris*	**Cholic**, allocholic, chenodeoxycholic, deoxycholic, lithocholic and other acids	Taurine	Yousef et al., 1977
Varanidae	*Varanus monitor*	$3\alpha,7\alpha,12\alpha,24\alpha$- and $3\alpha,7\alpha,12\alpha,24\beta$-tetrahydroxy-$5\beta$-cholestan-26(or 27)-oic acids; $3\alpha,7\alpha,12\alpha$-trihydroxy-5β-cholest-23-en-26(or 27)-oic acid; cholic and other acids	Taurine	Ali, 1975
SERPENTES				
Viperinae	*Bitis arietans; B. nasicornis; Vipera russelli; V. berus*	$3\alpha,7\alpha,12\alpha,23\xi$-tetrahydroxy-$5\beta$-cholanic, cholic, allocholic, bitocholic, chenodeoxycholic, deoxycholic and $3\alpha,7\beta,12\alpha$-trihydroxy-5β-cholanic acids	Taurine	Ikawa and Tammar, 1976

APPENDIX (continued)

Higher animal groups	Species	Bile alcohols; bile acids*	Type of conjugate	References
Birds				
RHEIFORMES	*Rhea americana*	Cholic, chenodeoxycholic and deoxycholic acids; **other acids** (63%)	Taurine	Tammar, 1970
CASUARIIFORMES	Emu, *Dromiceius novae-hollandiae*	**Cholic, chenodeoxycholic** and deoxycholic acids; other acids (12%)	Taurine	
GAVIFORMES	Loons, *Gavia* sp.	Cholic acid; chenodeoxycholic acid		Ripatti and Sidorov, 1973
SPHENISCIFORMES	Penguins:			U.o.
	Adelie, *Pygoscelis adeliae*	**Cholic, allocholic and chenodeoxy-cholic** acids	Taurine	
	Ringed, *P. antarctica*	**Cholic, allocholic, chenodeoxy-cholic** and deoxycholic acids	Taurine	
	Gentoo, *P. papua*	**Cholic, allocholic** and chenodeoxycholic acids	Taurine	
	Black-footed (Jackass), *Spheniscus demersus* and *S. humboldti*	**Cholic, allocholic, chenodeoxy-cholic** and deoxycholic acids	Taurine	
PROCELLARIIFORMES	Giant petrel, *Ossifraga gigantea*	**Cholic, allocholic and chenodeoxy-cholic** acids	Taurine	Tamar, 1970
PELICANIFORMES	Gannet, *Sula bassana*	**Cholic and chenodeoxycholic** acids; other acids (23%)	Taurine	
CICONIIFORMES	Shoebill stork, *Balaeniceps rex*	Cholic, chenodeoxycholic and deoxycholic acids; **other acids** (84%)	Taurine	

APPENDIX (continued)

Higher animal groups	Species	Bile alcohols; bile acids*	Type of conjugate	References
	Flamingoes: Giant, *Phoenicopterus antiquorum* and Chilean, *P. chilensis* Lesser, *P. minor*	Cholic, allocholic, chenodeoxycholic and other acids (30 and 7% respectively)	Taurine	Tammar, 1970
		Cholic, chenodeoxycholic and other acids (4%)	Taurine	
ANSERIFORMES	Horned Screamer, *Palamedea cornuta*	Chenodeoxycholic and other acids (15%)	Taurine	
	Snow goose, *Anser rossi*	Cholic, chenodeoxycholic and other acids (6%)	Taurine	
	Anser erythropus	Cholic acid (trace), chenodeoxycholic acid		
	Ducks: Pintail, *Anas acuta*; Widgeon, *A. penelope*; Mallard, *A. platyrhynchos*; Scaups, *Aythya sp.*; Goldeneye, *Clangula clangula*; Scoter, *Melanitta nigra* Black-necked swan, *Cygnus melancoriphus*	Cholic acid; chenodeoxycholic acid; other acid (s)		Ripatti and Sidorov, 1973
		Cholic and allocholic acids (trace); chenodeoxycholic and other acids (8%)	Taurine	
FALCONIFORMES	S. American vulture, *Catharista urubu*	Cholic, allocholic, chenodeoxycholic, deoxycholic and other acids (8%)	Taurine	Tammar, 1970
	Kite, *Milvus l. lineatus*	Trihydroxycoprostanic, $3\alpha,7\alpha$-dihydroxy-5β-cholestan-26(or 27)-oic, cholic and chenodeoxycholic acids	Taurine	Kuramoto and Hoshita, 1972

APPENDIX (continued)

Higher animal group	Species	Bile alcohols; bile acids*	Type of conjugate	References
	Lammergeier, *Gypaetus barbatus*	**Cholic**, chenodeoxycholic and deoxycholic acids	Taurine	
	African Sea eagle, *Cuncuma vocifer* and Bataleur eagle, *Tetrathopius ecaudatus*	**Cholic**, allocholic, chenodeoxycholic, deoxycholic and other acids	Taurine	
GALLIFORMES	Curassows, *Crax alector* and *Mitu mitu*	Chenodeoxycholic and other acids (70 and 36% respectively)	Taurine	Tammar, 1970
	Peafowl, *Pavo cristatus*	Chenodeoxycholic acid	Taurine	
	Green peafowl, *Pavo muticus*	Cholic, allocholic and **chenodeoxycholic** acids	Taurine	
	Domestic chick (germ-free)	Cholic, allocholic and **chenodeoxycholic** acids	Taurine	Haslewood, 1971
GRUIFORMES	Crowned crane *Balaerica pavonina*	Cholic and **chenodeoxycholic** acids	Taurine	Tammar, 1970
CHARADRIIFORMES	Golden plover, *Charadrius apricarius*	**Cholic acid**; chenodeoxycholic acid		Ripatti and Sidorov, 1973
	Herring gull, *Larus argentatus*	**Cholic acid**; chenodeoxycholic acid		
	Common gull, *Larus canis*	Cholic acid; chenodeoxycholic acid		
STRIGIFORMES	Eagle owl, *Bubo bubo*	Cholic, chenodeoxycholic and other acids (8%)	Taurine	Tammar, 1970
CORACIFORMES	Black thighed hornbill, *Bycanistes cylindricus*	Unidentified acids	Taurine	
PASSERIFORMES	Jackdaw, *Coloeus monedula*	**Cholic acid**; chenodeoxycholic and **chenodeoxycholic** acid		Ripatti & Sidorov, 1973
	Raven, *Corvus corone*	Cholic acid; **chenodeoxycholic** acid		
Mammals MARSUPALIA	Bennett's wallaby, *Macropus rufogriseus*	**Cholic**, chenodeoxycholic and deoxycholic acids	Taurine	Tammar, 1970
EUTHERIA INSECTIVORA	*Solenodon paradoxus*	**Cholic**, chenodeoxycholic and 3α,12α-dihydroxy-7-oxo-5β-cholanic acids	Taurine; glycine (trace)	U.o.[4]

APPENDIX (continued)

Higher animal group	Species	Bile alcohols; bile acids*	Type of conjugate	References
PRIMATES				
Tupaiidae	Tree "shrew", *Tupaia belangeri*	**Cholic, chenodeoxycholic,** deoxycholic, ursodeoxycholic and lithocholic acids		Schwaier, 1977, 1978
Cebidae	Squirrel monkey, *Saimiri sciureus*	**Cholic, chenodeoxycholic,** deoxycholic and lithocholic acids		Jones et al., 1976; Tanaka et al., 1976
Cercopithecidae	*Macaca irus*	**Cholic**, allocholic, chenodeoxycholic and deoxycholic acids	Taurine	Tammar, 1970
	Macaca maurus	**Cholic**, allocholic, chenodeoxycholic and **deoxycholic acids**	Taurine; glycine	
	Baboons, *Papio anubis*, *P. cynocephalus* and *P. ursinus*	Cholic, chenodeoxycholic, deoxycholic and lithocholic acids	Taurine; glycine	Kritchevsky et al., 1974; Morrisey et al., 1975
Pongidae	Orang utan, *Simia satyrus*	Cholic, chenodeoxycholic and **deoxycholic acids**	Taurine; glycine	Tammar, 1970
Hominidae	Man	3β,7α-Dihydroxy-4-cholen-24-oic and 3β,7α-dihydroxy-5-cholen-24-oic acids		Harano et al., 1976 and see Chapter 5
EDENTATA				
Myrmecophagidae	Anteater, *Myrmecophaga tridactyla*	**Cholic**, allocholic chenodeoxycholic and deoxycholic acids	Taurine	
Bradypodidae	Sloth, *Choloepus hoffmani*	Cholic and other acids	Taurine; glycine	
PHOLIDOTA				
Manidae	Pangolins *Manis pentadactyla* and *M. tricuspis*	**Cholic**, chenodeoxycholic and deoxycholic acids	Taurine	Tammar, 1970
LAGOMORPHA				
Leporidae	Jack rabbit, *Lepus californicus*	Cholic acid; **deoxycholic acid**	Taurine; glycine	Ripatti and Sidorov, 1973
	Hare, *Lepus timidus*	Cholic acid; dihydroxycholanic acids		
	Laboratory rabbit (germ-free)	**Cholic**, allocholic and chenodeoxycholic acids	Glycine	Hofmann et al., 1969

APPENDIX (continued)

Higher animal group	Species	Bile alcohols; bile acids*	Type of conjugate	References
RODENTIA				
Sciuridae	Ground squirrels *Ammospermophilus leucurus* and *Spermophilus tereticaudus*	Cholic, chenodeoxycholic, deoxycholic, $3\beta,12\alpha$-dihydroxy-5β-cholanic and lithocholic acids		Yousef et al., 1973
	Prairie dog, *Cynomys ludovicianus*	Cholic, chenodeoxycholic, deoxycholic and lithocholic acids		Den Besten et al., 1974
Heteromyidae	Pocket mouse, *Perognathus formosus*; Kangaroo rats *Dipodomys deserti*, *D. merriami* and *D. microps*	Cholic, chenodeoxycholic, deoxycholic, ursodeoxycholic, $3\beta,12\alpha$-dihydroxy-5β-cholanic and lithocholic acids		Yousef et al., 1973
Castoridae	Beaver, *Castor canadensis*	Cholic, **chenodeoxycholic** and deoxycholic acids	Taurine; glycine (mainly)	Tammar, 1970
Cricetidae	Pack rat, *Neotoma lepida*	**Cholic**, chenodeoxycholic and deoxycholic acids		Yousef et al., 1973
	Golden hamster, *Mesocricetus auratus*	Cholic, chenodeoxycholic and deoxycholic acids	Taurine; glycine	Dam, 1969
	Mongolian gerbil, *Meriones unguiculatus*	**Cholic**, chenodeoxycholic and deoxycholic acids		Noll et al., 1972
Muridae	Laboratory mouse	**Cholic, **allocholic, **α-, **β- and ω-muricholic, **chenodeoxycholic, deoxycholic, lithocholic and other acids (** in germ-free mice; β-muricholic is the chief bile acid in these animals)	Taurine; sulphate	Eyssen et al., 1976
	Laboratory rat (germ-free)	As mouse, but also allo(5α)chenodeoxycholic and 3β-hydroxy-5-cholen-24-oic acids	Taurine; sulphate	Eyssen et al., 1977 Gustafsson et al., 1977

APPENDIX (continued)

Higher animal groups	Species	Bile alcohols; bile acids*	Type of conjugate	References
Capromyidae	Hutias: *Capromys pilorides*	Cholic, 3α,12α-dihydroxy-7-oxo-5β-cholanic deoxycholic and 3α-hydroxy-7-oxo-5β-cholanic acids	Taurine (trace); glycine	Ripatti and Siderov, 1973
	Geocapromys browni	Cholic, 3α,12α-dihydroxy-7-oxo-5β-cholanic, chenodeoxycholic and 3α-hydroxy-7-oxo-5β-cholanic acids; other acids (traces)	Taurine; glycine (mainly)	U.o.[4]
	G. ingrahami	As *G. browni*, plus traces of deoxycholic and hyodeoxycholic acids		
	Plagiodontia aedium	Cholic, 3α,12α-dihydroxy-7-oxo-β-cholanic and deoxycholic acids	Taurine (trace); glycine	
CARNIVORA Canidae	Wolf, *Canis lupus*	Cholic acid; dihydroxycholanic acids		Ripatti and Siderov, 1973
	Maned wolf, *Chrysocyon brachyurus*	Cholic, allocholic, chenodeoxycholic and deoxycholic acids	Taurine; glycine	Tammar, 1970
Ursidae	Japanese and Himalayan bears	Cholic, chenodeoxycholic, deoxycholic and ursodeoxycholic acid	Taurine; glycine	See p. 111
	Himalayan bear, *Selenarctos tibetanus*	Cholic, chenodeoxycholic and deoxycholic acids	Taurine	
	American bear, *Ursus americanus*	Cholic and chenodeoxycholic acids	Taurine	
	Polar bear, *Thalarctos maritimus*	Cholic, chenodeoxycholic, deoxycholic and (?) ursodeoxycholic acids	Taurine	
	Malay bear, *Melarctos malayanus* and Sloth bear, *Melursus ursinus*	Cholic, chenodeoxycholic and deoxycholic acids	Taurine (Malay bear)	Tammar, 1970
Procyonidae	Racoon, *Procyon lotor*	Cholic, chenodeoxycholic and deoxycholic acids	Taurine	

APPENDIX (continued)

Higher animal group	Species	Bile alcohols; bile acids*	Type of conjugate	References
Mustelidae	Mink, *Mustela vison*; Marten, *Martes martes*; Otter, *lutra lutra*	Cholic acid; dihydroxycholanic acids	Taurine	Ripatti and Sidorov, 1973
	Fisher, *Martes pennanti*	Cholic, deoxycholic and chenodeoxycholic acids	Taurine	
	Sumatran otter, *Lutra sumatrans*	Cholic, 3α,12α-dihydroxy-7-oxo-5β-cholanic, chenodeoxycholic and deoxycholic acids; ?trace of allocholic acid	Taurine; ?glycine (trace)	U.o.[4]
	Sea otter, *Enhydra lutris*	As *L. sumatrans*, without allocholic acid	Taurine	
	Indian otter, *Amblonyx cinerea*	Cholic, chenodeoxycholic and/or deoxycholic acids	Taurine	Tammar, 1970
Viverridae	Mongoose, *Herpestes edwardsi*	Cholic, chenodeoxycholic and deoxycholic acids	Taurine	Tammar, 1970
Felidae	Puma, *Felis concolor*, Lion, *F. leo*, Leopard, *F. pardus*, *F. tigris* and Cheetah, *Acinonyx jubatus*	Cholic acid, allocholic acid (*F. concolor*); chenodeoxycholic and deoxycholic acids	Taurine (?glycine in *A. jubatus*)	Ripatti and Sidorov, 1973
	Felis lynx	Cholic acid; dihydroxycholanic acids		
	Domestic cat	Cholic, allocholic, chenodeoxycholic, deoxycholic, lithocholic, other acids	Taurine, sulphate	Taylor, 1977
PINNIPEDIA Otariidae and Phocidae	Californian sea-lion, *Zalophus californicus* and seal, *Phoca vitulina*	Cholic, chenodeoxycholic, deoxycholic and other acids	Taurine	Tammar, 1970
TUBILIDENTATA	Aardvark, *Orycteropus afer*	Cholic, allocholic, chenodeoxycholic and deoxycholic acids	Taurine	
PERISSODACTYLA	Domestic horse	Cholic and chenodeoxycholic acids	Taurine, glycine	Anwer et al., 1975
ARTIODACTYLA Suidae	Domestic pig (germ-free)	Cholic acid (probably); hyocholic, chenodeoxycholic and hyodeoxycholic acids	Taurine; glycine	Haslewood, 1971

APPENDIX (continued)

Higher animal group	Species	Bile alcohols; bile acids*	Type of conjugate	References
	Wart-hog, *Phacochoerus aethiopicus*	Cholic, chenodeoxycholic, deoxycholic and hyodeoxycholic acids	Taurine; glycine	Tammar, 1970
Hippopotamidae	*Hippopotamus amphibius*	Cholic, chenodeoxycholic, deoxycholic and other acids	Taurine; glycine (mainly)	
Bovidae	Kudus, *Strepsiceros kudu* and *S. imberbis*; Cape buffalo, *Bubalus caffer*	**Cholic**, chenodeoxycholic and deoxycholic acids	Taurine; glycine (*S. kudu* and *B. caffer*)	
	Gemsbok, *Oryx beisa*	Cholic, allocholic, chenodeoxycholic and deoxycholic acids		
	Blesbok, *Damaliscus albifrons*, Oribi, *Ourebia ourebi*, and Gerenuk, *Lithocranius walleri*	**Cholic**, allocholic, and deoxycholic acids	Taurine, glycine	Tammar, 1970
	Gazelles, *Gazella bennetti* and *G. rufifrons*; Tahr, *Hemitragus jemlahicus*	**Cholic**, chenodeoxycholic and deoxycholic acids	Taurine; glycine	
	Grecian wild goat and Mouflon, *Ovis musimon*	Cholic, allocholic, chenodeoxycholic and deoxycholic acids	Taurine and glycine (Mouflon)	
	Domestic sheep and lambs	**Cholic**, (a little unconjugated), chenodeoxycholic and deoxycholic acids	Taurine; glycine	Peric-Golia and Socic, 1968

*Chief bile acids or alcohols where these are known, are printed in **bold type**; where percentages are given these are the authors' estimates of percentages of the total bile alcohols or bile acids.

U.o. = Unpublished observations in the author's laboratory.
U.o.[1] = Unpublished results from the Zaire River Expedition 1974–1975. J. Linnean Soc. (In press)
U.o.[2] = Unpublished work with R. S. Oldham.
U.o.[3] = Unpublished work with T. Briggs.
U.o.[4] = Unpublished observations by Professor M. L. Johnson (University of Puget Sound, Tacoma, Washington, U.S.A.), Dr. I. G. Anderson, Dr. L. Tőkés and his colleagues and the author.

References Appendix

Ali, S. S. (1975) Bile composition of four species of amphibians and reptiles. Fed. Proc., 34, 660.

Ali, S. S., Farhat, H. and Elliott, W. H. (1976) Bile acids. XLIX. Allocholic acid, the major bile acid of *Uromastix hardwickii*. J. Lipid. Res., 17, 21—24.

Amos, B., Anderson, I. G., Haslewood, G. A. D. and Tökés, L. (1977) Bile salts of the lungfishes *Lepidosiren*, *Neoceratodus* and *Protopterus* and those of the coelacanth *Latimeria chalumnae* Smith. Biochem. J., 161, 201—204.

Anderson, I. G., Haslewood, G. A. D., Oldham, R. S., Amos, B. and Tökés, L. (1974) A more detailed study of bile salt evolution, including techniques for small-scale identification and their application to amphibian biles. Biochem. J., 141, 485—494.

Anwer, M. S., Gronwall, R. R., Engelking, L. R., and Klentz, R. D. (1975) Bile acid kinetics and bile secretion in the pony. Am. J. Physiol., 229, 592—597.

Briggs, T. and Bussjaeger, C. (1972) Allocholic acid, the major component in bile from the river Carpsucker, *Carpoides carpio* (Rafinesque) (Catostomidae). Comp. Biochem. Physiol., 42B, 493—496.

Bussjaeger, C. (1973) A comparative study of bile salts in sucker fishes (family Catostomidae). Ph.D. Thesis, University of Oklahoma.

Dam, H. (1969) Nutritional aspects of gallstone formation with particular reference to alimentary production of gallstones in laboratory animals. World Review of Nutrition and Dietetics, Vol. II, pp. 199—239. Karger, Basel—New York.

Denton, J. E., Yousef, M. K., Yousef, I. M. and Kuksis, A. (1974) Bile acid composition of Rainbow trout, *Salmo gairdneri*. Lipids, 9, 945—951.

DenBesten, L., Safaie-Shizaz, S., Connor, W. E. and Bell, S. (1974) Early changes in bile composition and gallstone formation induced by a high cholesterol diet in prairie dogs. Gastroenterology, 66, 1036—1045.

Eyssen, H. J., Parmentier, G. G. and Mertens, J A. (1976) Sulfated bile acids in germ-free and conventional mice. Eur. J. Biochem., 66, 507—514.

Eyssen, H., Smets, L., Parmentier, G. and Janssen, G. (1977) Sex-linked differences in bile acid metabolism of germfree rats. Life Sci., 21, 707—712.

Greenwood, P. H. (1975) A History of Fishes. J. R. Norman. 3rd Edn. Ernest Benn, London.

Gustafsson, B. E., Angelin, B., Einarsson, K. and Gustafsson, J-Å. (1977) Effects of cholesterol feeding on synthesis and metabolism of cholesterol and bile acids in germfree rats. J. Lipid Res., 18, 717—721.

Harano, K., Harano, T., Yamasaki, K. and Yosioka, D. (1976) Isolation of $3\beta,7\alpha$-dihydroxychol-5-en-24-oic and $3\beta,7\alpha$-dihydroxychol-4-en-24-oic acids from human bile. Proc. Japan. Acad., 52, 453—456.

Haslewood, G. A. D. (1967) Bile Salts. Methuen, London.

Haslewood, G. A. D. (1971) Bile salts of germ-free domestic fowl and pigs. Biochem. J., 123, 15—18.

Haslewood, G. A. D., Ikawa, S., Tökés, L. and Wong, D. (1978) Bile salts of Green turtle *Chelonia mydas* (L). Biochem. J., 171, 409—412.

Hofmann, A. F., Mosbach, E. H. and Sweeley, C. C. (1969) Bile acid composition of bile from germ-free rabbits, Biochim. Biophys. Acta, 176, 204—207.

Ikawa, S. and Tammar, A. R. (1976) Bile acids of snakes of the subfamily Viperinae and the biosynthesis of C-23 hydroxylated bile acids in liver homogenate fractions from the Adder, *Vipera berus* (Linn). Biochem. J., 153, 343—350.

Jones, D. C., Melchior, G. W. and Reeves, M. J. W. (1976) Quantitative analysis of individual bile acids by gas-liquid chromatography: an improved method. J. Lipid. Res., 17, 273—277.

Kellogg, T. F. (1975) The biliary bile acids of the Channel catfish, *Ictalurus punctatus* and the Blue catfish *Ictalurus furcatus*. Comp. Biochem. Physiol., 50B, 109—111.

Kihara, K., Yasuhara, M., Kuramoto, T. and Hoshita, T. (1977) New bile alcohols, 5α- and 5β-dermophols from amphibians. Tetrahedron Letters, 8, 687—690.

Kritchevsky, D., Davidson, L. M., Shapiro, I. L., Kim, H. K., Kitagawa, M., Malhotra, S., Nair, P. P., Clarkson, T. B., Bersohn, I. and Winter, P. A. D. (1974) Lipid metabolism and experimental atherosclerosis in baboons; influence of cholesterol-free, semi synthetic diets. Am. J. Clin. Nutr., 27, 29—50.

Kuramoto, T. and Hoshita, T. (1972) The identification of C_{27}-bile acids in kite bile. J. Biochem. Tokyo, 72, 199—201.

Kuramoto, T., Kikuchi, H., Sanemori, H. and Hoshita, T. (1973) Bile salts of anura. Chem. Pharm. Bull., 21, 952—959.

Morrisey, K. P., McSherry, C. K., Swarm, R. L., Nieman, W. H. and Dietrick, J. E. (1975) Toxicity of chenodeoxycholic acid in the nonhuman primate. Surgery, 77, 851—860.

Noll, B. W., Walsh, L. B., Doisy, E. A. Jr. and Elliott, W. H. (1972) Bile acids XXXV. Metabolism of 5α-cholestan-3β-ol in the Mongolian gerbil. J. Lipid. Res., 13, 71—77.

Noma, Y., Noma, Y., Kihara, K., Yashuhara, M., Kuramoto, T. and Hoshita, T. (1976) Isolation of new C_{26} alcohols from bullfrog bile. Chem. Pharm. Bull. 24, 2686—2691.

Okuda, K., Horning, M. G. and Horning, E. C. (1972) Isolation of a new bile acid, 3α,7α,12α-trihydroxy-5α-cholestan-26-oic acid, from lizard bile. J. Biochem. (Tokyo), 71, 885—890.

Peric-Golia, L. and Socic, H. (1968) Free bile acids in sheep. Comp. Biochem. Physiol., 26, 741—744.

Ripatti, P. O. and Sidorov, V. S. (1973) Quantitative composition of the bile acids of certain vertebrates in connection with the nature of their nutrition. Dok. Akad. Nauk. SSSR., 212, 770—773.

Schwaier, A. (1977, 1978) Die Verwendung von Tupaias als neues biologisches Testobjekt in der Präventivmedizin und angewandten medizinischen Forschung. Bericht fur das Bundesministerium für Forschung und Technologie, Bonn; and personal communication.

Tammar, A. R. (1970) A comparative study of steroids with special reference to bile salts. Ph.D. Thesis, University of London.

Tammar, A. R. (1974a, b, c) In: Chemical Zoology. Eds. M. Florkin and B. T. Scheer, a, Bile salts in fishes. Vol. VIII, pp. 595—612; b, Bile salts of Amphibia. Vol. IX, pp. 67—76; c, Bile salts in Reptilia, Vol. IX, pp. 337—351. Academic Press, San Francisco.

Tanaka, N., Portman, O. W. and Osuga, T. (1976) Effect of type of dietary fat, cholesterol and chenodeoxycholic acid on gallstone formation, bile acid kinetics and plasma lipids in Squirrel monkeys. J. Nutr., 106, 1123—1134.

Taylor, W. (1977) The bile acid composition of rabbit and cat gall-bladder bile. J. Steroid Biochem., 8, 1077—1084.

Yousef, I. M., Yousef, M. K. and Bradley, W. G. (1973) Bile acid composition in some desert rodents. Proc. Soc. Expl. Biol. Med., 143, 596—601.

Yousef, I. M., Bradley, W. G. and Yousef, M. K. (1977) Bile acid composition of some lizards from southwestern United States. Proc. Soc. Expl. Biol. Med., 154, 22—26.

Addendum

While this book was in proof, Dr. Herbert Falk held his Fifth Bile Acid Congress in Freiburg, West Germany; about 400 delegates attended and much new work was presented. Readers might like to hear of some of this and also of new discoveries recently reported. I refer to appropriate sections in the foregoing text.

1.5. Nature of bile pigments

F. Compernolle, K. P. M. Heirwegh and their colleagues (Compernolle et al., 1978) have shown that when aqueous solutions of 1-0-bilirubin-IXa-glucuronide (in which the bilirubin-IXa is attached by ester linkage to C-1 in glucuronic acid) are kept above pH 7, molecular rearrangements occur so that bilirubin becomes attached to C-2, C-3 or C-4 of the glucuronic acid; similar changes occur in cholestasis. This work explains some of the conflicting opinions about the smaller conjugated bilirubin molecules. Compernolle et al. could not find the bilirubin conjugates with other carbohydrates described by Kuenzle (1970).

3.1—3.4 Biosynthesis of bile acids and bile alcohols

Two undecided biochemical questions concerning bile acid biosynthesis are: (1) which of the two terminal methyl groups in the cholestane side-chain is oxidised to give the substances 3.7—3.11 in Scheme 3.1 (p. 50) and (2) whether oxidation is normally done by enzymes in the mitochondria or endoplasmic reticulum (microsomal fraction) of the liver cells.

When cholesterol is biosynthesised from ^{14}C-2-mevolonate (p. 68), one of the two terminal methyl groups in the side-chain becomes radioactive. This is the *pro*-R methyl group and G. Popják and his colleagues (1977) have now proposed that its carbon atom be numbered C-26. Popják et al. found it magnetically distinguishable from C-27 in cholesterol biosynthesised from ^{13}C-mevalonate.

J. Gustafsson and S. Sjöstedt (1978), using the microsomal fraction plus supernatant from liver cells of male Sprague—Dawley rats, converted cholesterol biosynthesised from ^{14}C-2-mevolonate into 5β-cholestane-3α,7α,12α-triol (Scheme 3.1) and thence into 3α,7α,12α-trihydroxy-5β-cholestan-26-oic acid. When this acid was decarboxylated, the radioactivity in the derived CO_2 was about 80% of that calculated to be on C-26 in the 5β-cholestane-3α,7α,12α-triol: thus the microsomal hydroxylase system was specific for C-26, as defined above. The authors, taking into account other published work, concluded that in normal bile acid biosynthesis in rats this hydroxylase system might be more important than the mitochondrial ω-hydroxylase which has the opposite stereospecificity and, unlike the microsomal system, hydroxylates cholesterol itself at C-27.

E.H. Mosbach and his co-workers, on the other hand (Dayal et al., 1978 and Fifth Bile Acid Congress, Freiburg) prepared 25R and 25S 5β-cholestane-3α,7α,26-triols and used these as standards to investigate the stereospecificity of terminal side-chain hydroxylation of 5β-cholestane-3α,7α-diol (Scheme 3.1) by subcellular fractions from livers of guinea pigs, rats, rabbits and human subjects. In all four species, Mosbach et al. found that ω-hydroxylation by the microsomal fractions was only 6—20% of the total hydroxylation, whereas with mitochondria ω-hydroxylation accounted for 65—85% of the total. Both C-25 isomers of 5β-cholestane-3α,7α,26-triol were produced in similar proportions by the microsomal fraction, but hepatic mitochondria made mainly the 25R triol suggesting that, at least in the biosynthesis of chenodeoxycholic acid, the initial side-chain hydroxylation is carried out by the mitochondria.

These conflicting views will no doubt be reconciled by future experiments. Biochemists and others will also be grateful for a clear final definition of which of the two side-chain terminal methyl groups in the cholesterol molecule is to be designated C-26, for the definition of C-26 now proposed by Popják et al. is the opposite of the one on

p. 22, which follows the suggestion made by K. A. Mitropoulos & N. B. Myant (1965). The IUB—IUPAC Rules merely state that "if one of the two methyl groups attached to C-25 is substituted it is assigned the lower number (26)", which is no help to biochemists. The trihydroxycoprostanic acid usually obtained from biles is the 25R isomer and 5α- and 5β-chimaerol have the same configuration (Anderson and Haslewood, 1970).

P. G. Killenberg and his colleagues have continued to elucidate the enzymic mechanism of bile acid conjugation. They have examined the subcellular distribution of the two enzymes (cholyl-CoA-synthetase: step 1, p. 66) and bile acid-CoA-amino acid N-acyltransferase: step 2, p. 66) that catalyse the formation of glycine and taurine conjugates of C_{24} bile acids. After partial purification of the second enzyme, the authors conclude that the rat probably has a single acyltransferase that uses either taurine or glycine (Killenberg, 1978; Killenberg and Jordan, 1978).

G. Parmentier and H. Eyssen (1977; 1978) have now described about 30 of the 39 possible sulphate esters of the three common bile acids, cholic, chenodeoxycholic and deoxycholic acids, and their glycine and taurine conjugates, together with other similar compounds. These standards will be of great value in studies of this interesting aspect of bile salt metabolism.

5.7. The gallstone problem

M. C. Carey and D. M. Small (1978) have carried out a systematic phase equilibrium study which showed that the total bile "lipid" (bile salt + lecithin + cholesterol) concentration is more important in determining the solubility of cholesterol than the bile salt-lecithin ratio. For example, at a fixed lecithin—bile salt mole fraction of 0.25 (a common figure for bile) an increase in total lipid concentration from 0.35 g/dl (very dilute hepatic bile) to 25 g/dl (very concentrated gallbladder bile) resulted in a sixfold increase in the molar concentration of cholesterol. The authors have worked out tables for the calculation of "percent cholesterol saturation" at various physiologically likely total lipid, bile salt and lecithin concentrations. At the Fifth Bile Acid Congress, Carey pointed out that for usual gall-bladder bile with a total lipid concentration of, say, 5—15 g/dl, calculation of the saturation index by the equations of Thomas and Hofmann (p. 134) might not result in errors of clinical importance; however, for hepatic bile these equations are quite inapplicable. The "micellar zone" proposed by Carey and Small is smaller still than that shown on Figure 5.1 and the new calculations show that "all cholesterol gallstone patients have supersaturated gallbladder biles". The new work may supersede the calculations given on p. 134.

For gall-bladder bile containing much ursodeoxycholic acid, which will become common in treated gallstone patients, a correction must be applied by subtracting a figure from the mole percent cholesterol values, since cholesterol is less soluble in mixtures of ursodeoxycholic acid taurine and glycine conjugates than in corresponding solutions of conjugates of chenodeoxycholic acid.

There appeared to be complete agreement at the Freiburg Congress that ursodeoxycholic acid is superior to chenodeoxycholic acid for the treatment of radiolucent cholesterol gallstones. It is effective in smaller doses, gives rise to less lithocholic acid and is free from the side effects (diarrhoea, raised serum transaminase) often noticed during treatment with chenodeoxycholic acid. Japanese medicine had an initial advantage in testing this therapy, for ursodeoxycholic acid has been available to its doctors for many years. Clinical trials of urso-deoxycholic acid and also of chenodeoxycholic acid are likely to continue and the metabolism and mode of action of the former compound will receive concentrated attention. The precise mechanisms by which these dihydroxy bile acids achieve a lowering of the saturation index of bile in vivo are by no means understood.

T. Low-Beer made the potentially useful observation that cholesterol solubility in bile in vivo could be increased by decreasing the concentration of deoxycholic acid conjugates, which can be done by the administration of substances preventing the growth of anaerobic intestinal bacteria responsible for the 7-deoxygenation of cholic acid conjugates. In 11 men given the drug "metronidazole" (selectively bactericidal to anaerobes) the proportion of deoxycholic acid conjugates in the gall-bladder bile fell considerably and in almost all subjects the saturation index also fell, rising when treatment ceased.

It was pointed out on p. 135 that chenodeoxycholic and ursodeoxycholic acids are biochemically connected by the 7-deoxy compound, 3α-hydroxy-7-oxo-5β-cholanic acid (7-ketolithocholic acid). The last substance can readily be made from chenodeoxycholic acid by the several types of intestinal bacteria having 7α-hydroxysteroid dehydrogenase activity. H. Fromm, A. R. Hofmann and their colleagues reported at Freiburg that 7-ketolithocholic acid during one passage through the human liver is reduced mainly to chenodeoxycholic and to a lesser extent to ursodeoxycholic acid. Thus it is clear how chenodeoxycholic acid can be converted in vivo to ursodeoxycholic acid; what is not so obvious is how the opposite change, i.e. ursodeoxycholic → chenodeoxycholic acid, can occur. There are no known common intestinal bacteria that have 7β-hydroxysteroid dehydrogenase activity, although of course such might yet be discoverd. Unless ursodeoxycholic acid is dehydrogenated at C-7, it is hard to see how chenodeoxycholic acid can arise from it except perhaps by some hitherto unknown epimerisation process, possibly in the liver.

Much other interesting work was described at Freiburg, for example I. A. Macdonald and co-workers described the discovery of a potentially very useful 12α-hydroxysteroid dehydrogenase preparation and A. Schwaier and H. Weis demonstrated the advantages of the tree "shrew" *Tupaia belangeri* as a laboratory animal for bile salt studies in relation to human disease; this primate has bile much more like that of man than other small laboratory species (see Appendix) and can easily be induced to form cholesterol gallstones. The proceedings of the Congress will be published in full by MTP Press, Lancaster, U.K. under the title "Biological Effects of Bile Acids: Falk Symposium No. 26".

References to Addendum

Anderson, I. G. and Haslewood, G. A. D. (1970) Comparative studies of bile salts. 5α-chimaerol, a new bile alcohol from the white sucker *Catostomus commersoni* Lacépède. Biochem. J. 116, 581—587.

Carey, M. C. and Small, D. M. (1978) The physical chemistry of cholesterol solubility in bile. Relationship to gallstone formation and dissolution in man. J. Clin. Invest. 61, 998—1026.

Compernolle, F. Van Hees, G. P., Blanckaert, N. and Heirwegh, K. P. M. (1978) Glucuronic acid conjugates of bilirubin-IXa in normal bile compared with post-obstructive bile. Transformation of the 1-0-acylglucuronide into 2- ,3- and 4-0-acyl glucuronides etc. Biochem. J., 171, 185—214.

Dayal, B., Batta, A. K., Shefer, S., Tint, G. S., Salen, G. and Mosbach, E. H. (1978) Preparation of 24(R) and 24(S)-5β-cholestane-3α,7α,24-triols and 25(R) and 25(S)-5β-cholestane-3α,7α,26-triols by a hydroboration procedure. J. Lipid Res., 19, 191—196.

Gustafsson, J. and Sjöstedt, S. (1978) On the stereospecificity of microsomal "26"-hydroxylation in bile acid biosynthesis. J. Biol. Chem., 253, 199—201.

Killenberg, P. G. (1978) Measurement and subcellular distribution of cholyl-CoA-synthetase and bile acid-CoA-aminoacid-N-acyltransferace activities in rat liver. J. Lipid Res., 19, 24—31.

Killenberg, P. G. and Jordan, J. T. (1978) Purification and characterization of bile acid CoA-aminoacid-N-acyltransferase from rat liver. J. Biol. Chem., 253, 1005—1010.

Mitropoulos, K. A. and Myant, N. B. (1965) Evidence that the oxidation of the side chain of cholesterol by liver mitochondria is stereospecific, and that the immediate product of cleavage is propionate. Biochem. J., 97, 26c—28c.

Parmentier, G. and Eyssen, H. (1977) Synthesis and characteristics of the specific monosulfates of chenodeoxycholate, deoxycholate and their taurine or glycine conjugates. Steroids, 30, 583—590. (1978) Thin-layer chromatography of bile salt sulphates. J. Chromat., 152, 285—289.

Popják, G., Edmond, J., Anet, F. A. L. and Easton, N. R. Jr. (1977) Carbon-13 NMR studies on cholesterol biosynthesized from [^{13}C]-mevalonates. J. Am. Chem. Soc., 99, 931—935.

Index

Species listed in the Appendix are not generally referred to in the Index.

Aadvark 113
Abrus precatorius 80
Absorption of fats 11, 141
  bile salts 3
  cholesterol 12
Acids $C_{27}H_{46}O_6$ 27
Adder 100
Alcohols in bile 24
Alkaline hydrolysis 37, 61, 95
Alligator mississipiensis 102
Allochenodeoxycholic acid 60
Allocholic acid 30
  biosynthesis 59
  occurrence 90, 98, 100, 103
  precursor 60
  properties 30
Allodeoxycholic acid 30
Allo(5α)lithocholic acid 136, 139
Amia calva 87
δ-Aminolaevulinate synthetase,
    affected by bile alcohols 13
Amniotic fluid 124
Amphibians 93 *et seq.*
Amphioxus 80
Analysis of bile salts, methods 35—39, 126
Amphiuma means 94
Anatomy of enterohepatic
    circulation 3, 4

Anurans 94—96
Aptenodites 102
Arapaima gigas 92
Arapaimic acid 27
Arapaimol-A, 24
  formula 92
  properties 24
Arapaimol-B 24
Archosauria 101
Arginine conjugates of bile acids 34
Arthropods and sterol biosynthesis 79
Artifacts of enterohepatic
    circulation 69

Baboons, bile salts in 131
Bacteria,
  effect of bile salts on 14
  effect on bile salts 69—71
  intestinal 71
Balaenoptera 110
Barbus 88
Beaver 109
Bile,
  canalicular 6
  caniculus 5
  cholesterol in 9—11, 128—130, 142
  collection 35
  composition 6, 122, 123
  ductular 6

enterohepatic circulation 3, 4
function of 6, 11—13
gall bladder 4, 122
hepatic 122
human 121—142
lipid 8—11
mucin 7
phospholipid 8, 9
pigments 7, 8, 185
protein 7
secretion 5
Bile acids 27, 30
analysis 35—39
biosynthesis 49—62, 185
C_{24} 30
C_{27} 27
conjugation 66
distribution in animals 79—115
in disease 136—142
faeces 32, 127
infants 124
serum 125
urine 127
primary 21, 59
properties 27, 30
secondary 21, 28, 69—71
sulphates of 34, 187
unconjugated 32
Bile pigments 7, 185
conjugated 8
human 8
in bile 8
Bile salts 21
absorption 3
analysis 35—39, 125
biosynthesis 49—69, 105—187
controlling biosynthesis 67
distribution in micelles 12, 130
effect on parasites 13
enterohepatic circulation 3, 4
evolutionary significance 79—115, 153—158
functions 11—13
human pool 135, 140
in Aardvark 113
amphibians 93—96
Artiodactyla 113—115
biliary atresia 139

birds 102, 103
Carnivora 111, 112
cerebrotendinous xanthomatosis 53, 140
Cetacea 110
cholestasis 136—138
chondrichthyans 82
Chondrostei 86
cirrhosis 138
coeliac disease 140
Crocodilia 101
Crossopterygii 83
cyclostomes 80
cystic fibrosis 139
Dipnoi 84
diverticulosis 140
Edentata 107
elephants 113
faeces 127
'Holostei' 87
human bile 123, 124
horses 113
infants 124
Insectivora 106
invertebrates 79
Lagomorpha 107
liver disease 136—139
mammals 103—115
marsupials 105
meconium 124
monotremes 105
normal blood 125
normal urine 127
Ostarophysi 88—91
Pholidota 107
pigs 113—115
Pinnipedia 112
Polypterus 87
primates 106
reptiles 96—102
Rodentia 108—110
Serpentes 99—101
serum 125, 136
stagnant loop syndrome 141
Teleostei 88—93
Testudines 96
urine in disease 137
Wilson's disease 140

molecular models 41, 42
physical chemistry 39
primary 21, 59
relation to diet, 104, 112, 128
secondary 21, 69—71
Biliary atresia 139
Bilirubin 7
biliary 8, 185
conjugated 8
glucuronide 8
macromolecular 7
Biliverdin 7
Biosynthesis 49—67, 185—187
control of 67
of allocholic acid 59
bile alcohols 62
chenodeoxycholic acid 51—57
cholic acid 49—51, 55
Birds 102, 103
26,27-Bisnor-5α-cholestane-3α,7α,12α,24-tetrol: see 26-deoxy-26-nor-5α-ranol
Bitis arietans 100
Bitis gabonica 100
Bitocholic acid 100
Boidae 99
Bombina orientalis, bile acids 61
Bran 135, 142
Branchiostoma lanceolatum 80
Breath test 141
Bufo b. formosus 33, 62, 95, 96
5α-Bufol 24
formula 94
properties 24
5β-Bufol 24, 62
Bullfrog 65, 95
Bycanistes cylindricus 103

Calamoichthys 87
Calcium,
in bile 122
gallstones 131
Cancer pagurus 79
Carnivora 111, 112
Carpiodes carpio 89
Castor canadensis 109
Catostomidae 89
Catostomus 89

Cats 111
Cattle 115, 130
Cebus 106
Cercopithicus 106
Cerebrotendinous xanthomatosis 53, 140
Cetacea 110
Chelonia mydas 26, 27, 96
Chenodeoxycholic acid 30
biosynthesis 51—57, 186
conversion to cholic acid 57
conversion to lithocholic acid 132
formula 56, 132
gallstone therapy 131—135
in blood 126
in human bile 123
properties 30
sulphates 34
5α-Chimaerol 24
formula 90
properties 24
5β-Chimaerol 24
formula 82
properties 24
Chimpanzee 133
Cholane 22
5α-Cholane-3α,7α,12α,24-tetrol: see 5α petromyzonal
5β-Cholane-3α,7α,12α,24-tetrol: see 5β-petromyzonal
Cholanic acid 29
Cholan-24-oic acid 29
Cholecystokinin 3
Choleic acids 140
Cholestane 21
5β-Cholestane-3α,7α-diol 50
5α-Cholestane-3α,7α, 12α,25,26,27-hexol: see 5α-dermophol
5β-Cholestane-2β,3α,7α,12α,26,27-hexol: see arapaimol-B
5β-Cholestane-3α,7α,12α,24,26,27-hexol: see scymnol
5β-Cholestane-3α,7α,12α,25,26,27-hexol: see 5β-dermophol
5α-Cholestane-3α,7α,12α,24,26-pentol: see 5α-chimaerol
5α-Cholestane-3α,7α,12α,25,26-pentol: see 5α-bufol

5α-Cholestane-3α,7α,12α,26,27-pentol:
see 5α-cyprinol
5α-Cholestane-3β,7α,12α, 26,27-pentol:
see latimerol
5β-Cholestane-2β,3α,7α,12α,26-pentol:
see arapaimol-A
5β-Cholestane-3α,7α,12α,23,25-
pentol 24, 53
5β-Cholestane-3α,7α,12α,24,25-
pentol 24, 53, 54
5β-Cholestane-3α,7α,12α,24,26-pentol:
see 5β-chimaerol
5β-Cholestane-3α,7α,12α,25,26-
pentol 53, see also 5β-bufol
5β-Cholestane-3α,7α,12α,26,27-pentol:
see 5β-cyprinol
5α-Cholestane-3β,7α,16α,26-tetrol: see
myxinol
5α-Cholestane-3α,7α,12α,26-tetrol: see
27-deoxy-5β-cyprinol
5β-Cholestane-3α,7α,12α,25-tetrol 53
5β-Cholestane-3α,7α,12α,26-tetrol:
see 27-deoxy-5β-cyprinol
5α-Cholestane-3α,7α,12α-triol 60
5α-Cholestane-3β,7α,26-triol 60,
see also 16-deoxymyxinol
5β-Cholestane-3α,7α,12α-triol 50
5β-Cholestane-3α,7α,26-triol 52, 136
5α-Cholestan-3β-ol 10, 60, 140
5β-Cholestan-3β-ol: see coprosterol
Cholestasis 8, 138
Cholesterol,
  biosynthesis 67
  conversion to bile acids 49—63,
    185—187
  estimation 10
  formula 50
  7α-hydroxylase 52
  in gallstones 130—135
  solubility in bile 128—130
Cholestyramine 137
Cholic acid,
  biosynthesis 49, 53
  conjugation 34, 66, 187
  conversion to deoxycholic acid 28,
    70
  formula 28
  free, in bile 33, 90
  in blood 126

  monoglucuronide 137
  properties 30
  sulphates 34, 187
Choloepus hoffmani 107
Chondrostei 86
Ciliatocholate 34
Cirrhosis 138
Clostridium perfringens 32
Cod, 12α-hydroxylation 57
Coelacanth 83
Coeliac disease 140
Colubridae 101
Conjugation 34, 66, 187
Control of bile salt biosynthesis 67
Coprosterol (coprostanol) 10
Corralus enhydris 99
Coypu 110
Crab 79
Crayfish 79
Crocodilia 101
Crossopterygi 83
Curassow 103
Cutting-grass 110
Cyclostomes 80
Cylindrophis 99
Cypriniformes 88—91
5α-Cyprinol,
  formula 84
  properties 24
5β-Cyprinol,
  formula 87
  properties 24
Cyprinus carpio 64
Cystic fibrosis 139

Decanoylsarcosyltaurine 79
25-Deoxy-5α-bufol 24, 89
24-Deoxy-5α-chimaerol 24, 89
Deoxycholic acid,
  formation from cholic acid 28, 70,
    188
  formula 29
  in blood 126
    enteroliths 140
    faeces 128
  properties 30
27-Deoxy-5α-cyprinol,
  formula 89
  properties 24

27-Deoxy-5β-cyprinol 25
16-Deoxymyxinol 25
26-Deoxy-26-nor-5α-ranol 25
26-Deoxy-5α-ranol 25
26-Deoxy-5β-ranol 25, 65
Dermophis mexicanus 93
5α- and 5β-Dermophols 25, 93, 94
Dictyocaulus 14
Diemyctylus 94
Digestion of fat 11
Digitonin 10
3α-7α-Dihydroxy-5α-cholanic acid: see allochenodeoxycholic acid
3α,12α-Dihydroxy-5α-cholanic acid: see lagodeoxycholic acid
3α,6α-Dihydroxy-5β-cholanic acid: see hyodeoxycholic acid
3α,6β-Dihydroxy-5β-cholanic acid 31
3β,6α-Dihydroxy-5β-cholanic acid 31
3α,7α-Dihydroxy-5β-cholanic acid: see chenodeoxycholic acid
3α,7β-Dihydroxy-5β-cholanic acid: see ursodeoxycholic acid
3β,7α-Dihydroxy-5β-cholanic acid 136
3α,12α-Dihydroxy-5β-cholanic acid: see deoxycholic acid
3β,7α-Dihydroxy-4-cholen-24-oic acid 75, 177
3α,7α-Dihydroxy-5-cholen-24-oic acid 57, 177
3α,7α-Dihydroxy-5β-cholestanoic acid 27, 61
7α,12α-Dihydroxy-4-cholesten-3 one 50, 156
3α,7α-Dihydroxy-12-oxo-5β-cholanic acid 136
3α,12α-Dihydroxy-7-oxo-5β-cholanic acid 71
Diverticulosis 140
Dogs 104, 130
Domestic fowl 103
Duodenal bile, composition 123

Echidna 105
Echinococcus granulosus, effect of bile salts on 13
Edentata 107
Eel, 12α-hydroxylation in 57

Electrophorus electricus 90
Elephas maximus 113
Endothelial cells 5
Enhydra lutris 112
Enterohepatic circulation 3, 4
Enteroliths 140
Enzymic analysis of bile salts 38, 125
Eptatretus 80
Erinaceus europaeus 106
Erpetoichthys 87
Escherichia coli, 7α-hydroxysteroid dehydrogenase in 38, 71, 105
Eutherians 106
Evolution and bile salts 79—115, 153—158

Faeces, bile salts in 32, 127
Fats,
 absorption 12
 digestion 11
 in bile 8—11, 122
Foetus, bile salts in 124

Gaboon viper 100
Gadus callarias 57
Gall-bladder 3
 anatomy 3, 4
 functions 4
 occurrence 3
Gallstones 130
 cholesterol in 131—135
 composition 131
 dissolution by chenodeoxycholic acid and ursodeoxycholic acid etc. 131—135, 187—189
 in Africans 131
  American Indians 131
  baboons 130
  cats 130
  ground squirrels 131
  guinea pigs 131
  hamsters 131
  mice 131
  pigs 130
  prairie dogs 131
  rats 130
  rabbits 131
  sheep 130

Geocapromys 110
Gerbil 60
Germ-free,
　animals 70
　mice 178
　pigs 114
　rabbits 70
　rats 178
Gigi fish 90
Glycine conjugates 29, 66, 104, 110
Glycochenodeoxycholate 123
Glycocholate 123
Glycodeoxycholate 123
Goats 115
Ground squirrel 131
G/T ratio 125, 138
Guinea pig 110

Haemulcholic acid 92, 166—168
Hagfishes 80
Hamster, gallstones in 109, 131
Hare 108
Hay's test 137
Heloderma 97
Henophidia 99
Hepatic bile 122
Hepatocyte 5
Herbivores 104, 115
Hippopotamus 115
Histidine conjugates of bile acids 34
HMG-CoA 67
"Holostei" 87
Hornbill 103
Horse, bile salts 113, 180
Human foetus and infant 124
Human bile,
　composition 122, 123
Human bile salts,
　composition 123
　estimation 125
　in bile 122, 123
　　faeces 127
　　liver disease 136—142
　　urine 127
　pool 135, 140
Hydrolysis of bile salts 32, 37
3α-Hydroxy-5α-cholanic acid 136, 139

3α-Hydroxy-5β-cholanic acid: see lithocholic acid
3β-Hydroxy-5-cholen-24-oic acid,
　biosynthesis 56
　formula 56
　in biliary atresia 139
　in infants 55, 124
3β-Hydroxy-5-cholesten-26-oic acid 56
7α-Hydroxy-4-cholesten-3-one 50
3-Hydroxy-3-methyl glutaryl CoA 67
3β-Hydroxycholenic acid: see 3β-hydroxy-5-cholen-24-oic acid
7α-Hydroxycholesterol,
　biosynthesis 52
　formula 50
26-Hydroxycholesterol 55, 56, 124, 136
6α-Hydroxylation 114
6β-Hydroxylation 109
7α-Hydroxylation
　of cholesterol 52
　of deoxycholic acid 71, 108
12α-Hydroxylation 57, 60, 61
23-Hydroxylation 59, 100
3α-Hydroxy-6-oxo-5α-cholanic acid 31, 115
3α-Hydroxy-7-oxo-5β-cholanic acid 105
　formula 105
　origin 70, 110
　properties 31
3α-Hydroxysteroid dehydrogenase 38, 125, 127
7α-Hydroxysteroid dehydrogenase 38, 70, 125, 127
Hymenolepis 14
Hyocholic acid
　biosynthesis 114
　formula 114
　properties 30
Hyodeoxycholic acid,
　biosynthesis 114
　formation in pig 114
　formula 114
　in wart-hog 115
　properties 31

Ictiobus 89
Iguana iguana 59, 173

Ileum 3, 4, 140, 142
Infant, bile salt biosynthesis 55, 124
Insectivora 106
Intestinal microorganisms,
　affected by bile salts 14
　effect on bile salts 21, 70, 71, 132
Invertebrates 79

Jaundice 8, 136—138
Jejunum 4, 140

Kite 103
Klebsiella pneumoniae 34
Koala 105
Kusimanse 111

Lacerta viridis 98
Lagodeoxycholic acid 31
Lagomorpha 107
Lampreys 81
Lagothrix 106
Latimeria chalumnae 83
Latimerol,
　formula 84
　properties 25
Lecithin,
　formula 9
　in bile 8, 129, 130
Lepidosiren paradoxa 85
Lepisosteus osseus 87
Lepus californicus 108
Lepus timidus 108
Liebermann-Burchard reaction 10
Lipase 11
Lipids,
　absorption 4, 11—13
　in bile 8—10, 128—130
　in micelles 12, 130
Lithocholic acid,
　biosynthesis 56
　formation from chenodeoxycholic
　　acid 132
　formula 132
　metabolism in man and other
　　primates 15, 34, 55, 124, 128, 133
　properties 31
　sulphation 133
　toxicity 131

Lithogenic index 133—135
Liver,
　canaliculi 5
　cells 5
　disease 136—139
　sinusoids 5
Lizards 97—100
Lungfish 84—86
Lutra sumatrans 112

Macaca mulatta 131
Mammals 103—115
Manis 107
Marsupials 105
Megalobatrachus japonicus 94
Meriones unguiculatus 60
Micelles 11, 12, 41, 129, 130
Microvilli 5, 12
Mitu mitu 103
Mongoose 111
Monitor lizards 97
Monkeys 106, 131
Monoglycerides 9, 11, 12, 122
Monosulphates of bile acids 34, 109,
　　124, 127, 133, 136, 187
Monotremes 105
Mucin 7, 130
α- and β- Muricholic acid,
　biosynthesis 56
　formulae 108
　properties 30
ω-Muricholic acid,
　biogenesis 109
　properties 30
Murinnae 108
Mus muscularis 108
Myocastor coypus 110
Myrmecophaga tridactyla 107
Myxine 80
Myxinol,
　disulphate, model 42
　formula 40
　properties 25

Necturus 94
Neoceratodus forsteri 85
27-Nor-5α-cholestane-3α,7α,12α,24,26-
　　pentol: see 5α-ranol

27-Nor-5β-cholestane-3α,7α,12α,24,26
   pentol: see 5β-ranol
27-Nor-5α-cholestane-3α,7α,12α,24-
   tetrol: see 26-deoxy-5α-ranol
27-Nor-5β-cholestane-3α,7α,12α,24-
   tetrol: see 26-deoxy-5β-ranol

Obstructive jaundice 137
Ophisaurus 98
Opossum 105
Ornithorhyncus anatinus 105
Orycteropus afer 113
Oryctolagus cuniculus 107
Ostariophysi 88—91
Osteichthyans 83—93

L-1-Palmityl-2-oleyl-phosphatidyl
   choline 9
Papio 131
Parapristipoma trilineatum 92
Parasilurus asotus 91
Parasites, effect of bile salts on 13
Pelteobagrus nudiceps 90
Penguins 102
Perrisodactyla 113
Petromyzon 81
5α-Petromyzonol,
   formula 81
   properties 25
5β-Petromyzonol 85
pH of bile 4, 121
Phascolarctos cinereus 105
Phocaecholic acid,
   formula 112
   occurrence 112
   properties 30
Phocochoerus aethiopicus 115
Phospholipids,
   in bile 8
   solubilising cholesterol 128—130, 187
Physical chemistry of bile salts 39—42
Pig,
   bile salts 113—115
   gallstones 130
Pinnipedia 112
Piranha 90
Plagiodontia aedium 110

Platynota 97
Platypus 105
Polypeptide-bilirubin complex 8
Polypterus 87
Porphyrin synthesis,
   effect of bile alchols on 13
Prairie dogs 131
Primary bile salts 21, 59
Primates,
   bile salts 106
   sulphation 133
Proboscidea 113
Procanbarus clarkii 79
Protein in bile 7
Protopterus aethiopicus 85
Pseudomonas testosteroni 38
Pythocholic acid,
   biogenesis 99
   formula 99
   properties 30
Rabbit,
   bile salt biosynthesis in 107
   germ-free 70
Radioimmunoassay 125, 126
Rainbow trout 167
Rana catesbeiana 64, 95, 171
5α-Ranol,
   formula 95
   properties 25
5β-Ranol,
   biosynthesis 65
   formula 65
   properties 25
Rats,
   bile salt biosynthesis in 70, 71, 108
   gallstones 130
Rattus 109
Reptiles 96—102
Rhesus monkey 131, 136
Rodentia 108—110
Rules of nomenclature 22, 29
Ruminants 9, 115

Saimiri 106
Salamander 61, 94
Salamandra salamandra 94
Sampoderma 98
Saturation index 133—135

Scyliorhinus canicula 83
Scymnol,
  formula 23
  history 23
  properties 25
Scymnus borealis 23
Sealions 112
Seals 112
Secondary bile salts 21, 69—71
Selenarctos 112
Serpentes 99
Serrasalmus ternetzi 90
Sheep 9, 34, 115
Siluriformes 90
Solubility of cholesterol in
  bile 128—135
Squamata 97
Stagnant loop syndrome 141
Sturgeons 86
Synodontis 91
Suidae 113
Sulphates,
  biosynthesis 66
  of bile acids 34, 35, 124, 127, 133, 136, 137, 187
  of lithocholic acid 34, 133
  solvolysis 37, 38
Sus scrofa 115

Tachyglossus aculeatus 105
Taurine conjugates,
  formation 66, 187
  of C_{24} acids 28
  of C_{27} acids 26
Taurochenodeoxycholate 123
Taurocholate,
  formula 40,
  model 41
Taurodeoxycholate 123
T/D ratio 123, 125, 138
Teleostei 88—93
Testudinata 96
$3\alpha,7\alpha,12\alpha,23$-Tetrahydroxy-$5\beta$-
  cholanic acid 30, 100, 101
$2\beta,3\alpha,7\alpha,12\alpha$-Tetrahydroxy-$5\beta$-
  cholestan-26 (or 27)-oic acid; see
  arapaimic acid
$3\alpha,7\alpha,12\alpha,22$-Tetrahydroxy-$5\beta$-
  cholestan-26 (or 27)-oic acid: see
  tetrahydroxysterocholanic acid
$3\alpha,7\alpha,12\alpha,24$-Tetrahydroxy-$5\beta$-
  cholestan-26(or 27)-oic acid: see
  varanic acid
Tetrahydroxyisosterocholanic acid 27
Tetrahydroxysterocholanic acid 27, 97
Thyronomys swindereanus 110
Tight junction 5
Tortoises 96
Triglycerides,
  absorption 12,
  digestion 11
  in bile 122
Trihydroxybufosterocholenic acid,
  formula 33
  origin 62
$3\alpha,7\alpha,12\alpha$-Trihydroxy-5α-cholanic
  acid: see allocholic acid
$3\alpha,6\alpha,7\alpha$-Trihydroxy-5β- cholanic acid:
  see hyocholic acid
$3\alpha,6\alpha,7\beta$-Trihydroxy-5β-cholanic acid:
  see ω-muricholic acid
$3\alpha 6\beta,7\alpha$-Trihydroxy-5β-cholanic acid:
  see α-muricholic acid
$3\alpha,6\beta,7\beta$-Trihydroxy-5β-cholanic acid:
  see β-muricholic
$3\alpha,7\alpha,12\alpha$-Trihydroxy-5β-cholanic
  acid: see cholic acid
$3\alpha,7\alpha,16\alpha$-Trihydroxy-5β-cholanic
  acid: see pythocholic acid
$3\alpha,7\alpha,22\alpha$-Trihydroxy-5β-cholanic
  acid: see haemulcholic acid
$3\alpha,7\alpha,23$-Trihydroxy-5β-cholanic acid:
  see phocaecholic acid
$3\alpha,12\alpha 23$-Trihydroxy-5β-cholanic
  acid: see bitocholic acid
$3\alpha,7\alpha,12\alpha$-Trihydroxy-5β-cholestan-26
  (or 27)-oic acid 61, 173
$3\alpha,7\alpha,12\alpha$-Trihydroxy-5β-cholestan-26-
  oic acid: see
  trihydroxycoprostanic acid
$3\alpha,7\alpha,12\alpha$-Trihydroxy-5β-cholest-22-
  ene-24-carboxylic acid: see
  trihydroxybufosterocholenic
  acid

3α,7α,12α-Trihydroxy-5β-cholest-23-
    en-26(or 27)-oic acid 62
Trihydroxycoprostanic acid,
    biosynthesis 50
    formula 50
    in human disease 139
    occurrence 61, 102, 169—172
Turtles 96
Typhlops 101

Unconjugated bile acids 32—34, 140
Urine,
    bile salts in 127, 136, 139
Urodela 93, 94
Uromastix hardwickii 61
Ursidae 111
Ursodeoxycholic acid,
    biogenesis 59, 70, 71, 188, 189
    formula 111
    gallstone therapy 135, 188
    properties 31
Ursus arctos 112

Varanic acid,
    formula 50
    in biosynthesis 50
    occurrence 61, 97
Varanidae 97
Varichorhinus 88, 167
Viper berus 100
Vipers 100
Viverridae 111

Walrus 112
Wart hog 115
Weasels 111
Whales 110
Wilson's disease 140

Xanthomatosis, cerebrotendinous 53, 140

Zaire River Expedition 88, 91, 181
Zwitterions 9